Science, Technology, and Society

A Sourcebook on Research and Practice

Edited by

David D. Kumar

Florida Atlantic University
Davie, Florida

and

Daryl E. Chubin

National Science Foundation
Arlington, Virginia

T0188848

Kluwer Academic/Plenum Publishers
New York, Boston, Dordrecht, London, Moscow

INNOVATIONS IN SCIENCE EDUCATION AND TECHNOLOGY

Series Editor:

Karen C. Cohen, Harvard University, Cambridge, Massachusetts

Are Schools Really Like This?: Factors Affecting Teacher Attitude toward School Improvement
J. Gary Lilyquist

The Hidden Curriculum—Faculty-Made Tests in Science
Part 1: Lower-Division Courses
Part 2: Upper-Division Courses
Sheila Tobias and Jacqueline Raphael

Internet Links for Science Education: Student-Scientist Partnerships
Edited by Karen C. Cohen

Web-Teaching: A Guide to Designing Interactive Teaching for the World Wide Web
David W. Brooks

Science, Technology, and Society
A Sourcebook on Research and Practice
Edited by David D. Kumar and Daryl E. Chubin

Time for Science Education
Michael R. Matthews

A Continuation Order Plan is available for this series. A continuation order will bring delivery of each new volume immediately upon publication. Volumes are billed only upon actual shipment. For further information please contact the publisher.

Science, Technology, and Society

A Sourcebook on Research and Practice

ISBN: 0-306-46173-0

©2000 Kluwer Academic/Plenum Publishers, New York
233 Spring Street, New York, New York 10013

http://www.wkap.nl

10 9 8 7 6 5 4 3 2 1

A C.I.P. record for this book is available from the Library of Congress

Printed in the United States of America

Contributors

Glen S. Aikenhead, Department of Curriculum Studies, College of Education, University of Saskatchewan, Saskatoon, SK S7N 0X1, Canada

James W. Altschuld, College of Education, The Ohio State University, Columbus, OH 42310

Dennis W. Cheek, Rhode Island Department of Education, Providence, RI 02903

Daryl E. Chubin, National Science Board Office, National Science Foundation, Arlington, VA 22230

Kathleen B. deBettencourt, Environmental Literacy Council, George Marshall Institute, Washington, D.C. 20006

Julie C. DeFalco, Competitive Enterprise Institute, Washington, D.C. 20036

Edward J. Hackett, Department of Sociology, Arizona State University, Tempe, AZ 85287

J. Scott Hauger, Research Competitiveness Program, American Association for the Advancement of Science, Washington, D.C. 20005

David Devraj Kumar, College of Education, Florida Atlantic University, Davie, FL 33314

Jon D. Miller, Center for Biomedical Communications, Northwestern University Medical School, Chicago, IL 60611-3008

Rustum Roy, Science, Technology, and Society Program, The Pennsylvania State University, University Park, PA 16802

Peter A. Rubba, Department of Curriculum and Instruction, The Pennsylvania State University, University Park, PA 16802

James A. Rye, Department of Educational Theory and Practice, West Virginia University, Morgantown, WV 26506

Preface to the Series

The mandate to expand and improve science education is an educational imperative and an enormous challenge. Implementing change, however, is very complicated given that science as well as science education is dynamic, continually incorporating new ideas, practices, and procedures; takes place in varying contexts; and must deal with amazingly rapid technological advances. Lacking clear paths for improvement, we can and should learn from the results of all types of science education, traditional as well as experimental. Successful reform of science education requires careful orchestration of a number of factors which take into account technological developments, cognitive development, societal impacts and relationships, organizational issues, impacts of standards and assessment, teacher preparation and enhancement, as well as advances in the scientific disciplines themselves. Understanding and dealing with such a complex mission is the focus of this book series. Each book in this series deals in depth with one or more of these factors, these potential factors for understanding, creating and sustaining effective science education improvement and reform.

In 1992, a multidisciplinary forum was launched for sharing the perspectives and research findings of the widest possible community of people involved in addressing the challenge. Those who had something to share regarding impacts on science education were invited to contribute. This forum was the *Journal of Science Education and Technology*. Since the inception of the journal, many articles have highlighted relevant themes and topics and expanded the context of understanding to include historical, current, and future perspectives in an increasingly global context. Recurring topics and themes have emerged as foci requiring expanded treatment and presentation. This book series, "Innovations in Science Education and Technology" is the result.

It is a privilege to be able to continue to elucidate and effect improvement and reform in science education by providing this in-depth forum for the works of others. The series brings focus and understanding to efforts worldwide, helping readers to understand, to incorporate, and to utilize what we know, what we are learning, and what we are inventing technologically to advance the mission of science education reform worldwide.

Karen C. Cohen
Cambridge, Massachusetts

Contents

INTRODUCTION

David D. Kumar and Daryl E. Chubin

We live in an information age. Technology abounds: information technology, communication technology, learning technology. As a once popular song went, "Something's happening here, but it's just not exactly clear." The world appears to be a smaller, less remote place. We live in it, but we are not necessarily closely tied to it. We lack a satisfactory understanding of it. So we are left with a paradox: In an information age, information alone will neither inform nor improve us as citizens nor our democracy, society, or institutions. No, improvement will take some effort. It is a heavy burden to be reflective, indeed analytical, and disciplined but only constructively constrained by different perspectives.

The science-based technology that makes for the complexity, controversy, and uncertainty of life sows the seeds of understanding in *Science, Technology, and Society*. STS, as it is known, encompasses a hybrid area of scholarship now nearly three decades old. As D. R. Sarewitz, a former geologist now congressional staffer and an author, put it

> After all, the important and often controversial policy dilemmas posed by issues such as nuclear energy, toxic waste disposal, global climate change, or biotechnology cannot be resolved by authoritative scientific knowledge; instead, they must involve a balancing of technical considerations with other criteria that are explicitly nonscientific: ethics, esthetics, equity, ideology. Trade-offs must be made in light of inevitable uncertainties (Sarewitz, 1996, p. 182).

Motivated by these concerns we created this book to contribute to classroom practice in STS across grade levels and subjects. We sought to stimulate debate on the future of STS, including where it fits on the intellectual landscape of the 21st century. Above all, we tried to suggest what citizenship will require and how STS is a well-lit, unending path to it. We present an assortment of content, advice, and examples. This is a book about teaching and learning and is designed for teaching and learning. A decade ago one of us wrote, "The value of studying science and technology as social phenomena is to learn about ourselves as we probe the natural world we inhabit" (Chubin and Chu, 1989, p. x). This remains compelling for us as editors. We hope it finds kindred spirits in our readers.

WHY STS?

Science and its social context continue to change. We are citizens of an increasingly global society influenced by an explosion of knowledge, advances in technology, and a progressive expansion of the free market. The hybrid character of STS mirrors this changing scene. STS draws on a range of intellectual sources: scientists and engineers seeking more than textbook treatments; educators focused on content that matches pedagogy; social scientists who insist that "context" imbues the science and technology with values, politics, and consequences. To philosophers, scientists, and commentators, science has become increasingly fragmented and compartmentalized, declining in its social relevance in academic circles. Edgar Morin (1998), for example, envisions for science an understanding of the "human condition" in the social milieu—too often a factor overlooked in the academy.

If time has seen STS programs institutionalized in many colleges and universities, it has diversified them as well. Some STS programs focus on values, ethics, business, environment, policy, etc. Like ethnic studies and women's studies programs, STS is a window for looking at the social and natural world differently. Its intellectual value stems from its breadth and its attentiveness to context and stakeholders in the outcomes of issues, controversies, and disputes that contain a science or technology component. Such value is reflected in STS curricula: The content of courses adds value through a perspective students lack when they enter college. Most students are grounded in a discipline and therefore are "content" specialists. Few, however, know how to view and analyze that technical content in a broader social, cultural, and policy context. If skepticism is an analytical virtue, then STS is a virtuous undertaking.

This collection recognizes the diversity of needs, of students, of classrooms served under the STS banner. Indeed, we as editors come from different worlds. One of us resides in a university and specializes in the role of STS curriculum in the teaching of science, while the other inhabits the policy world of a federal agency dedicated to integrating science education and research.

Science, Technology, and Society: A Sourcebook on Research and Practice offers a comprehensive view of STS as it enters the 21st century. The collection includes scholars in STS from both inside and outside the academic world. The goals of STS include: making science and technology literacy available for all; preparing the noncollege-bound student to compete successfully in a science- and technology-oriented workplace; and equipping the future citizenry with the tools and information necessary for making informed personal and policy decisions concerning the role of science and technology in global society.

STS, however, inhabits a bifurcated world. The evolution of STS in higher education has largely disconnected it from the community of scholars in the K–12 sector and the integration of STS themes with classroom pedagogy, especially in the sciences. Framed by the debates over national standards and the school-to-work transition, precollege STS is a different and separate scholarly pursuit from that of 20 years ago. Thus, this book seeks to unite scholars and their concerns for K–12 vs. higher education in a more seamless web of practice. Education specialists, K–12 classroom teachers, discipline-oriented scientists, mathematicians, and engineers must be able to understand each other and work together. Their shared experience can enrich STS and enable it to influence instruction and produce new generations of "knowledge workers": classroom teachers, researchers, and mentors, who are committed to drawing on the resources of an international interdiscipline and extending insights through STS studies.

This book also examines how STS affords perspectives on the role of science and technology in society that single disciplines miss or overinterpret. The future requires the application of a new integration of knowledge to appreciate and solve social problems with science and technology as a component. Contributors to this book were selected for their breadth of knowledge and integrative talents. They see the "big picture" and they seek to portray its nuances to student and nascent professional audiences who demand new tools, skills, and approaches to teaching and comprehending a complex reality.

Environment, technology, ethics, policy, regulation, and curriculum are but a few of the issues that also serve as tools, resources, and research sites for the STS educator. This collection includes works by experienced STS scholars who practice in diverse cultures and work settings. The chapters summarize literature and point the way to fruitful investigations in and out of the classroom by those dedicated to nurturing new science and technology-engaged citizens and workers.

Finally, this collection is not about a static body of knowledge producers who respect intellectual disciplines and know how to apply concepts and techniques to challenging problems. Instead, it is about an open-ended opportunity to debate STS from unconventional and contemporary viewpoints. It is in this spirit that "STS" refers not only to teaching and learning, but also to mentoring human resources for American society—its work force and within it, the sectors of government and educational institutions.

What Is the Instructor's Overarching Purpose?

While we cannot answer this question, we can declare our intentions as editors: We fancied this collection as a resource for the teaching and learning of STS. Specifically, we hope to demonstrate: the vitality of STS as a perspective and approach; documented insights that augment and challenge, if not correct, impressions left by traditional science instruction; and the existence of STS at the K–12 level and STS as undergraduate and graduate specializations. Admittedly, this last reason is more esoteric, but it has great potential impact. These communities are only tenuously connected and have only begun to be mutually supportive.

What Do the Chapters Offer?

STS should draw on the resources of various international interdisciplines. Just as STS research should not be divorced from STS pedagogy or activism, these chapters should be used as threads for weaving new patterns of scholarship for use in and out of classrooms. To this end, chapter authors were encouraged to include:

- A review of STS trends and issues from the literature that point the way for science-literate citizens and workers;
- A clear analysis of curriculum trends and issues in the segment of STS addressed (e.g., high school), including relevance to a particular substantive area such as environmental education;
- Applications of STS for teaching, learning, and mentoring situations;
- Policy implications of STS, including the level of government or particular agency with primary responsibility for action; and
- Issues for further research.

Given these injunctions, we as editors tried to read the book as we imagined our readers would, and we asked how the resulting chapters could be used in STS. Here are our candid suggestions, offered as an assist to adopters and would-be adopters alike.

Who Is the Audience?

Classroom teachers must decide whether they can use this collection for a background resource or as a text to be read by students. We would argue that the length of the chapters favors use by pedagogues, not novices. The writing assumes too much sophistication from high school students. This is both good and bad. It means that an instructor will have to adopt the collection as a whole or use only a selected subset of chapters. Here we provide a set of roadmaps that anticipates the kinds of courses and themes supported by different chapter subsets.

Survey and Review of STS Trends and Issues
Chapters 2–7
Advocacy–introspective Study
Chapters 1, 5, 7–10
Curriculum Trends and Issues
Chapters 2, 4–8
Teaching, Learning, and Mentoring Applications
Chapters 1–4, 6, 8, 11
Policy
Chapters 1–7, 9–11
Research
Chapters 5, 6, 9–11

How Is the Collection Organized?

Given that no one sequence of chapters will meet all classroom needs, we will characterize the authors and the contents, and make some observations that elaborate the themes above as possible configurations of use. Various combinations may serve an instructor's needs, depending on the level of the course, inclinations of the instructor, knowledge base–experience of the students, etc. As editors, we detect a mix of chapter types—by tone as well as content.

Several chapters are oriented more to higher education programs and audiences. Most of the authors are in higher education settings. The rest do STS in nonprofit policy research organizations and federal and state governments (where policy analysis is a common thread). Here is a synopsis of chapter themes:

In Chapter 1, Rustum Roy explains how the value of technology should be recognized as the pedagogical center of public understanding of policy, ethics, and society—from elementary grades through graduate study. He views STS as an approach with untapped potential. Dr. Roy is Evan Pugh Professor of the Solid State, professor of Science, Technology, and Society, and professor of Geochemistry at The Pennsylvania State University, as well as a leader in the STS movement in the United States. He is a member of the National Academy of Engineering (United States), and a foreign member of the Royal Swedish Academy of Engineering Sciences.

Chapter 2, by Jon D. Miller, presents the state of civic science and technology literacy in the United States as revealed by secondary analyses of various national databases. Several models focus on school-based learning at primary through college levels. These help to explain the origins and dimensions of the public's literacy in science and technology. Dr. Miller is director of the International Center for the Advancement of Scientific Literacy, and professor and director, center for Biomedical Communications at Northwestern University.

In Chapter 3, Glen S. Aikenhead illustrates how four processes lead to four general products in STS by providing successful examples of STS in Canada. The processes are deliberation on policy, research and development, implementation, and instructional assessment, and the products are curriculum policy, classroom materials, teacher understanding, and student learning. Dr. Aikenhead is professor of Science, Technology, and Society at the University of Saskatchewan, Canada.

Chapter 4, by Julie C. DeFalco, takes a legal and economic perspective in its analysis of science and technology regulatory policies. Examined are the Endangered Species Act, the Corporate Average Fuel Economy standards, and the Airbag Mandate. Ms. DeFalco presents a compelling argument for incorporating the unintended consequences of such science and technology policies in STS classrooms. She is an adjunct policy analyst at the Competitive Enterprise Institute.

Chapter 5, by James W. Altschuld and David Devraj Kumar, looks at STS as being supplementary and perhaps preferable to the typical teaching and learning of science. Also addressed are accountability–evaluation criteria in STS and the lack of sufficient STS themes in state science curriculum frameworks. The authors raise questions for research and development in STS. Dr. Altschuld is professor of Evaluation and Education Policy at The Ohio State University, and Dr. Kumar is professor of Science Education at

Florida Atlantic University and a Fellow of the American Institute of Chemists.

Chapter 6, by Kathleen B. deBettencourt, supplies the environmental education perspective on curriculum and policy issues related to STS. She uses the Independent Commission's content analysis of textbooks in the United States to systematically point out the lack of depth of science content in environmental education with implications for STS. Dr. deBettencourt is project director of the Independent Commission on Environmental Education at the George Marshall Institute in Washington, D.C.

Chapter 7, by Dennis W. Cheek, addresses the underrepresentation of technology in the K–12 STS paradigm and analyzes selected STS curriculum materials with a particular emphasis on the treatment of technology. Dr. Cheek is director of the Office of Information Services and Research at the Rhode Island Department of Education.

In Chapter 8, James A. Rye and Peter A. Rubba discuss students' pre- and post-instructional alternative conceptions of global warming as part of STS curricula and teaching for understanding. Dr. Rye is assistant professor of Science Education at West Virginia University, and Dr. Rubba is professor of Science Education and chair of the Department of Curriculum and Instruction at The Pennsylvania State University.

Chapter 9, by J. Scott Hauger, explores university-based STS programs. He outlines graduate approaches to STS, its disciplinary connections, and implications for both research and practice. Dr. Hauger is director of the Research Competitiveness Program at the American Association for the Advancement of Science.

Chapter 10, by Daryl E. Chubin, looks at the need for a new social compact for national science policy, and prescribes how STS-savvy policies can reshape thinking about federal funding, the status of universities, and the debunking of science. A version of this chapter appeared in "Science and Public Policy" (1996). Dr. Chubin is senior policy officer for National Science Board at the National Science Foundation.

In Chapter 11, Edward J. Hackett explores the history and contemporary state of STS research within the competitive organizational culture of the National Science Foundation. Dr. Hackett is professor of Sociology at Arizona State University. He is a former program director for Science and Technology Studies at the National Science Foundation.

We hope this collection of multiple perspectives in STS helps our readers become better informed about the role of science and technology in society. Only better teachers, researchers, mentors, and communicators will reform another generation of classroom practices, curricula, policies, and citizens in STS.

REFERENCES

Chubin, D. E., and Chu, E. W. (1989). *Science Off the Pedestal: Social Perspectives on Science and Technology*, Belmont, CA: Wadsworth Pub. Co.
Morin, E. (1998). Education: Reforme ou reformettes? *Le Monde*, June 18, 1998.
Sarewitz, D. R. (1996). *Frontiers of Illusion: Science, Technology, and the Politics of Progress*, Philadelphia: Temple University Press.

CHAPTER 1

Real Science Education: Replacing "PCB" with S(cience) through STS throughout All Levels of K–12
"Materials" as One Approach

Rustum Roy

INTRODUCTION

I write from the perspective of a chemist–materials researcher who has spent 50 years on the faculty of Pennsylvania State University. I founded its most recognized research unit, the Materials Research Laboratory, and directed it for a quarter century. I have published nearly 700 papers; over 40 were done in the last two years. I cite this to establish my credentials as a card-carrying, active research scientist.

Rustum Roy, STS Program, The Pennsylvania State University, University Park, PA 16802.

Science, Technology, and Society: A Sourcebook on Research and Practice, edited by Kumar and Chubin, Kluwer Academic / Plenum Publishers, New York, 2000.

1. What Went Wrong with "Science" Education? Mislabeling.

My involvement with STS goes back 20 years and my professional connection with "science" education goes back even farther. Right from the beginning, I recognized the sloppy use of language was the problem when science was discussed in the public forum. Everybody who talked about "public understanding of *science*" played on the (congressional) leaders' interest in *technology*. Thus, I received the first ever National Science Foundation project (INPUT) for Increasing The Public Understanding of Technology. Very early on I recognized *materials* technology was a particularly suitable and powerful access route to introduce, link, and connect people to the "science" they need.

2. Science ≠ Physics + Chemistry + Biology (P + C + B)

When pushed, every scientist, every science educator, every science student will admit that. Yet 50 million young people have the equation $S = P + C + B$ drummed into them (usually in high school) by the existence theorem that S in fact $= P + C + B$. Science is institutionalized in all American school systems as equal to physics, chemistry, and biology. Word has gotten around for generations about such science. Two words especially: *hard* and *dry* (and not connected to life). A third word, which does not get around, is the root of both: $P + C + B$ are (too) *abstract*. Many people simply cannot handle abstractions remote from life. Nor is there any reason for them to do so. But Western universities are descended from that monstrous absurdity raised to an Enlightenment mantra: *cogito ergo sum* (I think, therefore I am). Compounding the hubris of Aristotle in separating science for the elite from techne for the slaves, we thus created the fundamental problem associating the word science with abstract *thought*, with upper classes, and with Cartesian head–hand dichotomy.

THE OLD AND NEW PARADIGMS FOR LEARNING "SCIENCE"

In *Lost at the Frontier: U.S. Science Policy Adrift*, a book I co-authored with Deborah Shapley, we made the case against another gross error that has become enshrined as gospel in United States science policy and worse in the United States population in general. That is the so-called linear model of epistemology and making knowledge into reality.

$$BASIC\ ABSTRACT\ SCIENCE \xrightarrow{leads} APPLIED\ SCIENCES \xrightarrow{leads}$$
$$ENGINEERING \xrightarrow{leads} TECHNOLOGY$$

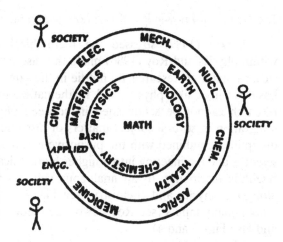

Figure 1. The last 40 years have been dominated by the basic science→applied science→ technology concept. Reprinted with permission from Shapley and Roy, 1985.

The empirically verifiable route is exactly the opposite, going from felt need→STS issues or problem→technology needed→applied science→ basic science (for a few that care).

Figures 1 and 2 illustrate the difference in another way. They present the changing paradigm more compellingly than words can.

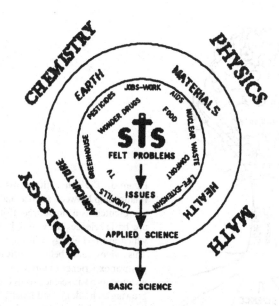

Figure 2. The new paradigm total reverses the theory illustrated in Figure 1, putting human concerns at the center and applied sciences (including materials) outside that to become the first science encountered. Reprinted with permission from Shapley and Roy, 1985.

The Nature of Knowledge of Technology and Science

Larkin (1989) stressed the hierarchical structure of knowledge within physics and Roy (1986) made the case that many applied sciences, such as materials research, do not lie in the same hierarchical plane as the basic sciences like physics and mathematics. In other words, materials research cannot be sandwiched in between physics and chemistry. The integration of several subject matters or disciplines, including engineering disciplines, combined with the purposive nature of the work, puts applied sciences and engineering into a higher hierarchical plane than the abstract scientific disciplines. In an analogous vein, technology is not a subject alongside physics and chemistry (see Fig. 2). It includes science as one among many inputs (see Roy's Two tree theory in Shapley and Roy, 1985 and also Figs. 3 and 4).

The idea that learning science is the necessary precursor to learning technology is absurd. As historians such as Derek de Solla Price and Melvin Kranzberg have long pointed out, all of human history is proof. Indeed the United States Department of Defense has shown that specific, even "high tech" tasks can be taught well, without any science. There are many entry points into the system of learning about technology. Figure 1 shows different routes that may be employed.

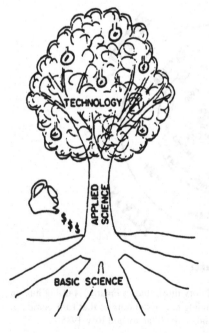

Figure 3. The conventional wisdom about the relationship between science and technology is represented as a single tree. If its roots, which are basic science, are watered, then the fruits, which are technology, will grow automatically. This widespread belief is disproved by the postwar experience of Great Britain (which has watered its basic science roots but has difficulty growing technology) and Japan (which has little basic science but grows technological fruits). Reprinted with permission from Shapley and Roy, 1985.

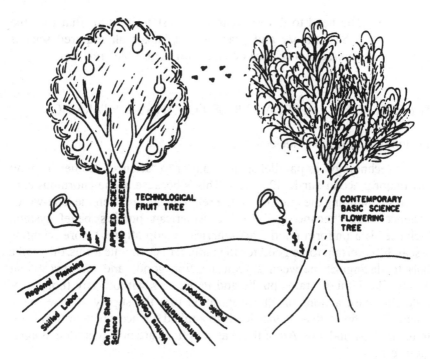

Figure 4. A more accurate metaphor is two trees: one for basic science and one for technology. This recognizes their distinct intrinsic character and that they are nurtured separately by separate policies. Each tree, when properly nourished, produces its own fruit; the basic science tree grows Nobel Prizes, and the more complex, applied science–engineering–technology tree grows technology. Reprinted with permission from Shapley and Roy, 1985.

For *the median learner* in K–12 we believe that the STS route—entering via the interest in the societal problem—is best. Moreover, it is the only innovation in *content* proposed for alleviation of the so-called math–science crisis. For about 10 percent of the population, entering via science (the present tradition in the United States) may be the most effective. But for a larger minority, the entry through hands-on technology may be the best. The United States has been losing out on the "brains in the fingertips" of the artisan, the "techne-ologist" by overstressing abstract conceptualization as the *only* way to learn technology itself and the science that is related to technology. The next section focuses on the new options.

If we accept this model of how to educate (nourish) 90 percent of our people in Technology, and hence the *Real* Science they need, we will need to address:

    a.   The need to "radically" abandon the P + C + B traditional approach.

  b. The need to devise a curriculum, state by state, that uses the
      S–STS theory and practice and makes all required science
      applied real science dealing with "sens-able" realities.

THE REAL SCIENCE APPROACH BASED ON THE S–STS PARADIGM

Technology Education Neglected

"Technology" as parallel subject matter to "science" has never found
any major place in our K–12 system. This is because of the enormous confusion surrounding the question of the relationship between the icon-words
"science" and "technology." In the American public's belief system,
"science" is a uniform good. The American credo affirms "more scientific
research" is certain to be good for the nation. In economic terms, the public
fails to distinguish between a "consumption good" and an "investment
good." The United States public and its leaders, without any thought or
reflection base actions on the proposition that the supply of new "basic
science" is infinite, that science leads to applied science which in turn leads
to technology and jobs. All of these assumptions are now regarded as egregious errors.

However, the United States attitude toward technology is much more
ambivalent. "High tech" carries the same cachet as "science"; but technology as polluter, negligent cause of adverse health effects (from war to
asbestos to "chemicals"), conjures up powerful negative images.

This situation was compounded by a mistaken belief that arose in
America after World War II. Victory was ascribed to the atomic bomb. Even
today only a tiny fraction of the population realize Japan had offered surrender before the bomb was dropped, and the atomic bomb was hailed and
celebrated as a product not of United States *technology*, but of physics!!
Thus was "science" ensconced in America's pantheon.

Finally, while "science" (now represented by P + C + B) became firmly
ensconced in the school system, vocational education, carrying many other
connotations, was the only toehold by which anything resembling "technology" gained influence within the school system. It is possible that
another historic shift will allow technology to be re-entered into mainstream K–12 education.

Loss of Comparative Advantage in Top Trained Personnel

The end of the second half of the 20th-century dominated by America
is now clearly in sight. The world of international technology has no clear

concentration of science and technology training or expertise. The United States still puts in large sums of taxpayer's money into abstract and esoteric research, in a level playing field devoted to research and development. The opportunity to return to a measure of reality will never be greater. The present United States "science emphasis" approach has been a devastating failure for United States technology and the economy, and this reality must be proclaimed and reinforced at every opportunity by anyone concerned about better technology education.

Opportunity and Responsibility

Those concerned with real science and technology education face an enormous challenge. First, they must clarify the relationships between science and technology—especially the place of both in the context of the economy and the political life of the country. Second, they must rethink, *de novo*, how and what one would teach the *average citizen* about technology, and what should be taught about science.

For the historical background and a discussion of what kinds of science and technology we need, the reader is referred to several recent papers (Roy, 1990; Roy 1992a, 1992b; Yager and Roy, 1993). We turn now to the new strategy that introduces the applied sciences to students. This includes materials science as the *primary* contact with science for at least 90 percent of students.

Strategy: Pedagogy from the Obvious, Instead of the Obscure

For centuries, communicating "techne" meant passing on from generation to generation important stored-up knowledge and wisdom about obvious, common, and often encountered human contacts with those parts of reality affecting humans the most. Each generation learned as much as possible about food, shelter, security, and so forth and passed it on to the next. For the last century, and over the last 50 years, school systems have attempted to teach *all* students *about* reality viewed from the particular formalism and stance of abstract science. This science is characterized by two key parameters: abstraction and mathematicization. These features are responsible for the power and rapid growth of science. They are at the same time responsible for its unintelligibility to, and lack of interest by, the vast majority of the population. Moreover, common sense and widespread human experience show that most people do *not* need much abstract science, and only modest quantification, to function very effectively, even in a highly technological society. The last president of the United States, the chairpersons of most of our largest corporations, the leading playwrights, poets, and university presidents have very little knowledge of the level of science some would demand of *all* students. It is not at all clear that a course in Advanced Placement physics or chemistry would have made an iota of difference to them.

A technology-focused curriculum would eschew abstraction for obviousness. Every one would be expected to know about the science of those parts of contemporary human experience that are obvious yet affect *all* in daily living.

A simple algorithm to guide the choice of what to know, which can expand and deepen with advancing grade simply by going into greater detail, is to follow the activities of an average pupil through an average day. From the alarm clock, to the light switch, to the clothes worn, to the rubber in the sneakers, to the stove that heats water for coffee, to the car that drives us to work, to the falling snow, and to the salting of highways, we have an infinite opportunity to take these objects and experiences and use them for teaching technology and applied science, and *derivatively*, basic science. This "applied science" must become the *necessary core* for all students, before they are exposed to *any* abstract science. The beauty of using the same common human experience—eating, getting dressed, driving—*is that they can be updated for* student's successive age level; and with increasing depth and sophistication, that can form the connecting introduction to any part of physics, chemistry, and biology. This is the technological literacy necessary for all; it is also much better groundwork for making science more attractive to many people.

The New Pedagogic Strategy: STS→Technology→Science

I believe student bodies that are exposed to STS will benefit in several ways:

1. They will be much more informed and aware of the most significant current issues involving science and technology.
2. They will have been shown a method of critically analyzing such issues.
3. They will have been made aware of how technology affects their lives, and how they may interact with technology.
4. A higher percentage than at present may choose to enter engineering—some may do so because they perceive it as a means of controlling their own futures.
5. A higher percentage will become interested in the scientific background behind the engineering, and this could result in more candidates for science degrees.

Thus the STS approach to "science" education has two separate benefits: better educated citizens and possibly increased enrollments in science and engineering.

Educating Americans in Real Science such as Materials Science

If the foregoing is an accurate, albeit necessarily qualitative and anecdotal description of the present situation of educating Americans about and in technology, it would call for several radical reforms in the entire structure and content of K–12 education in technology and real science. The major and substantive change should be in rectifying the gross and unnatural imbalance in all formal education toward abstraction and away from relevance and concreteness in all technical subject matter. This kind of change is necessary. This degree of abstraction from felt and experienced reality is what has isolated the entire culture of science and technology from most Americans. Science must be re-reified. Lemons and scrubbing ammonia must be the starting point for introducing pH; toasters and irons must lead through fuses to amps, volts, and watts.

While all the Real Sciences—agriculture, earth, engineering—may be intrinsically valuable paths to science knowledge, and also the pathways leading to standard P + C + B knowledge, the mother science of materials offers a special opportunity. Why? Materials are ubiquitous. Everyone encounters their local geology and also the weather, but not many are connected to agriculture. Health can be treated exactly as materials—a doorway to the biological world. But much of physics and chemistry, including the most recent work, is present and touchable and visible in new materials.

Special Role for Materials Science

I have used the materials available to every student in their classroom and on their way to their classes to link not only to the abstract and the esoteric materials science but also to STS. For example, starting with the clothing each student wears—cotton, wool, polyester mixtures—it is easy to develop the STS issues of global warming, resource depletion, polymer from oil, job shifting and offshoring, to the density of polymers, metal exhaustion, etc. In Richland, Washington, Steve Piipo, the pioneering physics teacher, has taken the local materials research known to all the tri-county area: radioactive waste forms; and connected it to all of physics and chemistry. By this connecting to real materials with real meaning in the life of the community, he has helped increase their enrollments in physics and chemistry dramatically.

The metals, plastics, and glasses every human being uses must be the seedbed from which the teaching of thermodynamics and the periodic table sprouts. Global climate issues daily reinforce the reality of the earth as a system from which can issue biodiversity, life forms, evolution, and so forth. Every illness, pill, and surgical procedure can serve as the "bait" for biology and lure another fraction of the students who have not responded to the abstract approach.

But, and this is of the utmost importance, it is not because more students may be enticed into entering or "appreciating" technology or science that this change must be made. It is much more fundamental than that. It is the repositioning and replacement of science back into its proper niche as one among many human activities, potentials, values, ideologies, and so forth. Moreover, it is this reconceptualization that will ultimately rescue basic science, which is quickly running out of things to study at a price the public (the only possible patron) is willing to pay. If science is not to become baroque, besides being broke, the bridges of the everyday world must be straightened. The replacement of the British–American Nobel Prize-dominated economies by the Japanese economy as the dominant economic force with its *technology-driven science* may bring home the point to the masses. Einstein once commented that if a culture's pipes did not hold water, neither would its theories. Yet thousands of graduate students in physics, chemistry, and even electrical engineering would be baffled by Einstein's claim of the close connection between our technology and our science because the reductionist paradigm has held that they can be paid from the public purse to do theoretical physics without any concern for their country's economic or technological base.

It is not appropriate here to develop and justify an optimum scope and sequence of the courses in science, technology, and STS, which could effectively educate the *median student*. An appropriate mix of K–12 teachers, education professors, and school administrators needs to be assembled to do just that. Yet, from the foregoing one can summarize some of the elements that should be present in any new curriculum for an STS and applied science approach to education of the median student. Table 1 covers some of the key content areas to be brought together under any such curriculum. Table 2 provides a rough sequence for educating American students in STS and *technology* and applied science.

Table 1. Key Items to be Included in New Curricula

1. Require STS components throughout 6–12.
   a. Distinction between science and technology. Relationship of science and technology to society: STS.
   b. Role of science and technology in the interaction of science, technology and global society.
2. Introduce formal science via applied science courses (materials, earth, health, agriculture).
3. Require some "technology" of every junior and senior high student along with their science requirements.
4. Shift emphasis of special programs from very science-talented to science-alienated (a fraction of whom are also talented).

Table 2. Possible STS and Technology Education Emphasis in the New Sequence

Grade	7	8	9	10	11	12
Science				Biology Chemistry Physics	Biology Chemistry Physics	Biology Chemistry Physics
Applied Science and Technology	Agriculture Science Concepts	Earth and Tech Ed	Materials and Tech Ed	Health and Tech Ed	Physics and Chemistry via Earth and Material	Physics and Chemistry via Earth and Material
STS	STS Critical Issues of the Day	STS Critical Issues of the Day	STS Problems	STS Unit Link to Appl. Sci.	STS Principles	STS Unit Link to Appl. Sci

REFERENCES

Larkin, J. (1989). Cognition in learning physics. *Amer. J. of Physics* 49(6):534–542.

Roy, R. (1986). Pedagogical theories and strategies for education in materials research. In Hobbs, L. (ed.), *Frontiers in Materials Education*, Materials Research Society, Pittsburgh, pp. 23–33.

Roy, R. (1990). The relationship of technology to science and teaching of technology. *J. of Tech. Edu.* 1(2):5–19.

Roy, R. (1992a). K–12 Education: A primer for material researchers. *MRS Bulletin* 5–9.

Roy, R. (1992b). Materials education: The second time around. *MRS Bulletin* 22–26.

Shapley, D., and Roy, R. (1985). *Lost at the Frontier—U. S. Science and Technology Policy Adrift*, ISI Press, Philadelphia, PA.

Yager, R. E., and Roy, R. (1993). STS most pervasive and most radical of reform approaches to science education in the STS movement. In The Science Technology Society Movement Yager, R. E. (ed.), NSTA Press, Washington, D.C. pp. 7–13.

CHAPTER 2

The Development of Civic Scientific Literacy in the United States

Jon D. Miller

Americans and other citizens of modern industrial societies live in an age of science and technology. Most adults in the industrialized world live in homes heated and cooled by a combination of thermostats and microcomputer chips. They watch pictures of world events transmitted by satellite unfold on their color television screen, and eat foods prepared and preserved by a wide array of technologies unknown to their parent's generation. When they become ill, they are treated with new pharmaceutical products that reflect 20th century advances in antibiotics, virology, or genetic engineering. For work, play, or family reasons, millions of Americans routinely take commercial air transportation to destinations around the planet.

Today's children—the next generation—will undoubtedly live in a significantly more scientific and technological culture. The rapid expansion of computer technology promises to relieve human beings of an ever larger

Jon D. Miller, Center for Biomedical Communications, Northwestern University Medical School, Chicago, IL 60611-3008.

Science, Technology, and Society: A Sourcebook on Research and Practice, edited by Kumar and Chubin, Kluwer Academic / Plenum Publishers, New York, 2000.

share of routine and repetitious work. New advances in agriculture and plant genetics suggest that the perennial struggle to feed the world's population will require less and less effort. Advances in medicine, communications, and transportation may lead to significantly longer lives and to a world community able to talk and visit with one another routinely. The curve of scientific and technological advance is still strongly positive.

The economic need for and value of a scientifically literate populace are well known. Science and technology have had a pervasive impact on both the methods of production and the products that are manufactured. The production of traditional industrial products like steel and the shaping of this and other metals into products has been largely automated. Work in the modern office is characterized by the machines and technologies utilized: word processors, data entry operators, database managers, fax clerks, and photocopy technicians. The industrial challenges of the 21st century will be the manufacture of microcomputer chips, genetically-engineered products, and new products yet to be invented. In this kind of economy, a basic understanding of science and technology will be the starting point for the development of the additional professional and technical skills needed to be competitive in an era of intense international economic competition.

Parallel to the need for a more scientifically literate workforce, the economy of the 21st century will need a higher proportion of scientifically literate consumers. From the experience of the last two decades, it is clear that increased exposure to computers at work and school has stimulated a strong and growing home microprocessor market. As more products incorporate new technologies, the information about the desirability, safety, and efficacy of those products will require a basic level of scientific literacy for comprehension. Some 20th century technologies, such as the irradiation of foods for preservation, have never achieved a high level of commercial success because of public misunderstanding and resistance. A strong technology-based economy in the 21st century will require that a substantial portion of the consuming populace be scientifically literate.

Of equal importance to these economic arguments, the preservation of democratic governments in the 21st century may depend on the expansion of the public understanding of science and technology. Over recent decades, the number of public policy controversies that require some scientific or technical knowledge for effective participation has been increasing. At the community level, the fluoridation controversies and referenda of the 1950s and 1960s in the United States illustrated the importance of a scientifically literate electorate. The more recent controversies over recombinant DNA field tests or proposed sites for nuclear power plants and nuclear waste disposal facilities point again to the need for an informed citizenry in the formulation of public policy. At the national level, the primary technological controversy of the latter decades of the 20th century has been the ongoing debate over the role of nuclear power in the production of energy. As the debate has widened to include the potential effects of continued burning of fossil fuels on planetary ecology, the need for a basic level of scientific literacy has become ever more urgent.

In the early decades of the 21st century, the national, state, and local political agenda will likely include an increasing number of important scientific and technological policy issues. While a more detailed discussion of public participation in the formulation of science and technology policy is beyond the scope of this chapter, it is important to note that the public is likely to play the role of final arbiter of disputes, especially when the scientific community and the political leadership are divided on a particular issue. As new energy and biological technologies move toward the marketplace, there will be important public policy issues to be decided. Some of these issues may erupt into full-scale public controversies. The preservation of the democratic process demands that there be a sufficient number of citizens able to understand the issues, deliberate the alternatives, and participate in the political processes leading to the resolution of public policy disputes involving science and technology.

If citizens are to discharge this responsibility in the context of an increasingly scientific society then a significant proportion of the electorate must be able to understand important public policy disputes involving science or technology. I refer to this level of understanding as *civic scientific literacy*. This chapter will propose a definition of civic scientific literacy, and describe the present level of civic scientific literacy in the United States, and explore the developmental origins of adult civic scientific literacy. The chapter will conclude with some suggestions concerning policies to enhance the level of civic scientific literacy in the United States.

THE CONCEPTUALIZATION AND MEASUREMENT OF CIVIC SCIENTIFIC LITERACY

The first basic conceptual issue concerns the scope of scientific literacy. Drawing from the basic concept of *literacy*, meaning the ability to read and write, scientific literacy might be defined as the ability to read and write about science and technology (Harman, 1970; Resnick and Resnick, 1977). But, given the wide array of scientific and technical applications in everyday life, scientific literacy might include everything from reading the label on a package of food, repairing an automobile, to reading about the newest images from the Hubble telescope. Approximately two decades ago,

Shen (1975) suggested that the public understanding of science and technology might be usefully divided into practical scientific literacy, cultural scientific literacy, and civic scientific literacy. In this context, *civic scientific literacy* refers to a level of understanding of scientific terms and constructs sufficient to read a daily newspaper or magazine and to understand the essence of competing arguments on a given dispute or controversy. Shen argued:

> Familiarity with science and awareness of its implications are not the same as the acquisition of scientific information for the solution of practical problems. In this respect civic science literacy differs fundamentally from practical science literacy, although there are areas where the two inevitably overlap. Compared with practical science literacy, the achievement of a functional level of civic science literacy is a more protracted endeavor. Yet, it is a job that sooner or later must be done, for as time goes on human events will become even more entwined in science, and science-related public issues in the future can only increase in number and in importance. Civic science literacy is a cornerstone of informed public policy. (p. 49)

Through her studies of the nuclear power controversy in Sweden, Nelkin (1977) has provided a useful framework for thinking about the content of civic scientific literacy. In the early 1970s, Sweden was seeking to develop a national policy on the use of nuclear power to generate electricity. To facilitate a broader public debate, the Swedish government provided small grants for *study circles* to discuss the nuclear power issue, usually in groups of 10–15 citizens with materials and a facilitator to provide a balanced presentation of the points of view. After months of discussions with approximately 80,000 Swedish citizens, the (Swedish) National Board of Civic Information conducted a study. They found that the portion of Swedish adults who felt unable to make a decision, having heard both arguments set forth, increased from 63 percent prior to the study circles to 73 percent after ten hours of study and discussion in study circles. Since it is primarily at the point of controversy that the public becomes involved in the resolution of scientific and technological disputes, it is clear that meaningful citizen participation requires a level of civic scientific literacy sufficient to understand the essential points of competing arguments and to evaluate or assess these arguments (Miller, 1983a).

Ziman and Wynne have attacked the basic idea of seeking to define and measure the understanding of scientific concepts, referring to this kind of analysis as a "deficit" model (Ziman, 1991; Wynne, 1991; Irwin and Wynne, 1996). They believe scientific meaning should be socially negotiated and that it should not be presumed that the knowledge of scientists is better than the common sense of nonscientists. Durant *et al.* (1992) have provided a thoughtful defense of the idea of defining and measuring public knowledge:

> We do not share Levy-Léblond's apparent willingness to divorce the ideals of democracy and literacy. On the contrary, we believe that the health functioning of democracy depends crucially upon the existence of a literate public; and in modern industrial societies, true democracy must embrace scientific literacy. (p. 163)

> ... there remains the problem of stigmatization. Clearly to measure levels of scientific understanding within a population is inevitably to assign higher scores to some individuals than others. By analogy with the notoriously controversial issue of IQ testing, this may be seen as inherently normative. Surely, it may be said, by measuring scientific understanding we are automatically branding as inferior those who score badly? Not at all. It is worth remembering that the French psychologist Alfred Binet developed the IQ test in order to identify those pupils who were most in need of educational assistance ... (demonstrating) that there is nothing necessarily prejudicial about the wish to find out how well individuals are doing in any particular area of educational and scientific attainment. (pp. 163–164)

In his discourse on the concept of scientific literacy, Shamos (1995) generally accepts the notion of a consumer scientific literacy and a civic scientific literacy, but, reflecting his own training in physics, insists on reserving the label of *true* scientific literacy for those who understand the third law of thermodynamics in essentially the same terms as a physicist. Although Shamos appears to accept the idea of civic scientific literacy at some points in his discourse, he ultimately concludes that citizens can never acquire sufficient understanding to participate in science and technology disputes, and embraces the long-discredited concept of a science court to remove science policy from the democratic process. Unable to step outside his own scientific training, Shamos fails to recognize that the general political institutions of society are extremely reluctant to exclude areas of decision-making from democratic influence, as shown in the uneasy experiment with independent regulatory commissions for securities, trade practices, and communications over the last four decades in the United States. Any effort to exclude science policy from the normal democratic processes would almost immediately foster similar demands for exclusive nondemocratic arrangements from numerous other interest groups.

Given the strong likelihood that science and technology policy will remain within the normal democratic policy formulation process in most countries, it is important to develop useable measures of civic scientific literacy to better understand its origins and its function in modern democratic systems. Building on a series of national surveys initiated in 1979, Miller (1983b, 1987a, 1995, 1997, 1998) has attempted to develop an empirical estimate of the proportion of American adults who qualify as being civic scientifically literate; his work remains the only empirical effort to provide an estimate of the proportion of adults qualifying as civic scientifically literate.

THE MEASUREMENT OF CIVIC SCIENTIFIC LITERACY

Miller has argued civic scientific literacy is a multidimensional construct. In a 1983 *Daedalus* article, Miller (1983b) suggested civic scientific literacy be conceptualized as involving three related dimensions: (1) a vocabulary of basic scientific constructs sufficient to read competing views in a newspaper or magazine; (2) an understanding of the process or nature of scientific inquiry; and (3) some level of understanding of the impact of science and technology on individuals and on society. It was argued that the combination of a reasonable level of achievement on each of these three dimensions would reflect a level of understanding and competence to comprehend and follow arguments about science and technology policy matters in the media. In more recent cross-national studies of civic scientific literacy, Miller found the third dimension—the impact of science and technology on individuals and society—to vary substantially in content among different nations and adopted a two-dimensional construct for use in cross-national analyses (Miller *et al.*, 1997).

In his early work, Durant recognized a two-dimensional structure for scientific understanding, but opted to use a continuous index of 27 items to measure the public understanding for analytic purposes, preferring to avoid use of the literacy concept and the establishment of a threshold that would classify individual respondents as literate and illiterate (Durant *et al.*, 1989, 1992; Evans and Durant, 1995). In more recent work, Durant and his colleagues suggested a three-dimensional model, but have continued to utilize only the vocabulary or construct understanding dimension for analysis (Bauer *et al.*, 1994).

Over the last 15 years there has been a growing agreement that civic scientific literacy can be usefully conceptualized as a two-dimensional measure, reflecting a vocabulary dimension of basic scientific constructs and a process or inquiry dimension. The desirability and feasibility of using a third dimension that reflects the social impact of science and technology in conceptualizing civic scientific literacy is still a point of some disagreement. There is general agreement among scholars engaged in national surveys, however, that a reliable two-dimensional measure of civic scientific literacy would be useful in a wide range of national and cross-national research.

Since 1972, the National Science Foundation has sponsored periodic surveys of public attitudes toward and understanding of science and technology. Beginning in 1979, the size and scope of the *Science and Engineering Indicators* surveys were expanded to include more attitudinal and knowledge items, and the number of open-ended inquiries used in these studies increased gradually throughout the 1980s and 1990s (National

Science Board, 1981, 1983, 1986, 1988, 1990, 1992, 1994, 1996, 1998). The 1991, 1993, 1996, and 1998 *Science and Engineering Indicators* studies and a 1993 Biomedical Literacy Study included an expanded set of knowledge items.

Using this database, a confirmatory factor analysis found two correlated, but analytically separable, factors that reflect the two dimensions described by Miller (Miller *et al.*, 1997; Miller, 1998). Nine knowledge items loaded on a construct, or vocabulary, dimension, while three other items defined a dimension reflecting an understanding of the nature of scientific inquiry (see Table 1). The nine items reflect the kinds of core concepts a citizen might need to understand a newspaper or magazine article, or a television report, concerning a scientific or technical issue. Given the limitations of time and respondent fatigue inherent in survey research, this set of items should be seen as a sampling of a larger universe of a hundred or more items that a well-informed citizen might need to comprehend and follow current science and technology policy issues. Item-response-theory (IRT) techniques were used to calibrate the nine items into a single dimension reflecting the extent of each individual's vocabulary of basic scientific constructs (Bock and Zimowski, 1997). Approximately 29 percent of American adults qualified as having a functional vocabulary of scientific constructs in 1997.

The second factor included three items, reflecting an understanding of the nature of scientific inquiry. When a citizen reads a news story or sees a television report about the results of a new medical experiment, for example, can that individual discern the difference between an appropriately designed and conducted experiment and vigorous medical claims without rigorous experimental testing, as in the Laetrile case in the United States? Although it is not reasonable to expect a scientifically literate citizen to be able to design or conduct an experiment, it is increasingly necessary for citizens and consumers to be able to recognize a scientific approach from a nonscientific or pseudoscientific approach. This three-item factor included one open-ended question concerning the meaning of studying something scientifically, one open-ended item asking for an explanation of the rationale for control groups, and a four-part item measuring the level of understanding of simple probability statements (see Table 1). Although only three items loaded on this dimension, each item involved either an open-ended explanation or a relatively complex multipart answer. Collectively, these three items provide a relatively rigorous test of the level of understanding of the nature of scientific inquiry.

The small number of items, however, cause some problems for the conversion of the responses into a single scale score. An alternative approach involves the construction of a typology. In previous work, Miller

Table 1. Confirmatory Factor Results, United States, All Adults, 1997

	Construct Knowledge Dimension	Process Knowledge Dimension	Proportion of Variance Explained
Provide a correct open-ended definition of a molecule.	.77	—	.60
Provide a correct open-ended definition of DNA.	.77	—	.59
Disagree that "Lasers work by focusing sound waves."	.72	—	.52
Indicate, through a pair of closed-ended questions, that the Earth goes around the Sun once each year.	.71	—	.51
Disagree that "All radioactivity is man-made."	.67	—	.45
Agree that "Electrons are smaller than atoms."	.59	—	.35
Indicate that light travels faster than sound.	.49	—	.28
Agree that "The continents on which we live have been moving their location for millions of years and will continue to move in the future."	.48	—	.23
Disagree that "The earliest humans lived at the same time as the dinosaurs."	.43	—	.19
Provide an open-ended explanation of the meaning of studying something scientifically.	—	.69	.47
Demonstrate an understanding of experimental logic by selecting a research design and explaining in an open-ended response the rationale for a control group.	—	.66	.44
Demonstrate an understanding of the meaning of the probability of one-in-four by applying this principle to an example of an inherited illness in four separate questions.	—	.54	.30

χ^2 = 99.1 / 51 degrees of freedom; Root Mean Square Error of Approximation = .022; Upper limit of the 90% confidence interval for RMSEA = .028; Correlation between Factor 1 and Factor 2 = .91; N = 2,000.

(1983b, 1987a, 1995, 1998) utilized a typology approach. Reflecting the underlying concept, a typology was constructed that classified all of those respondents who (1) were able to provide either a theory-building response to the scientific study question or a correct response to the experimental design question and (2) were able to define the meaning of probability as having a minimally acceptable level of understanding of the nature of scientific inquiry. Approximately 24 percent of American adults met this criterion in 1997.

As noted above, the two dimensions are positively and strongly correlated, but statistically and conceptually separable. Conceptually, individuals who demonstrate a high level of understanding on both dimensions are the most capable of acquiring and comprehending information about a science or technology policy controversy, and these individuals will be referred to as being "well informed" or "scientifically literate." At the same time, individuals who demonstrate either an adequate vocabulary of scientific constructs or who display an acceptable level of understanding of the nature of scientific inquiry are more capable of receiving and utilizing information about a science or technology policy dispute than other citizens who understand neither dimension. This second group will be referred to as "moderately well informed" or "partially scientifically literate." In the 1997 study, 15 percent of American adults qualified as well informed, or civic scientifically literate, and approximately 26 percent qualified as moderately well informed.

Over the last decade, the percentage of American adults qualifying as civic scientifically literate has increased from about 10 percent in 1988 to 15 percent in 1997 (see Fig. 1). While some portion of this increase can be attributed to a general increase in adult recognition of the importance of understanding basic scientific ideas and continuing improvements in the quantity and quality of informal science education resources in the United States, it is important to recognize that some part of this growth reflects the earlier experiences of these adults as students. To improve our understanding of the role of family, schools, and teachers in the development of civic scientific literacy and to gain some insight into the prospects for continued growth in civic scientific literacy, it is important to turn to the construction and examination of two models of the development of science achievement during the middle school and high school years and the growth of civic scientific literacy during the high school and college years.

THE DEVELOPMENT OF CIVIC SCIENTIFIC LITERACY

Parallel to the adult studies described in the preceding section, a longitudinal study of the development of student achievement in science was undertaken in the United States, providing a comparable database with designed points of linkage and comparison. The Longitudinal Study of American Youth (LSAY) was designed to study the development of student interest and competence in science and mathematics during the middle school, high school, and college years. The LSAY started tracking two cohorts of public school students in the fall semester of 1987. Throughout the United States approximately 3,000 seventh-grade students from 50

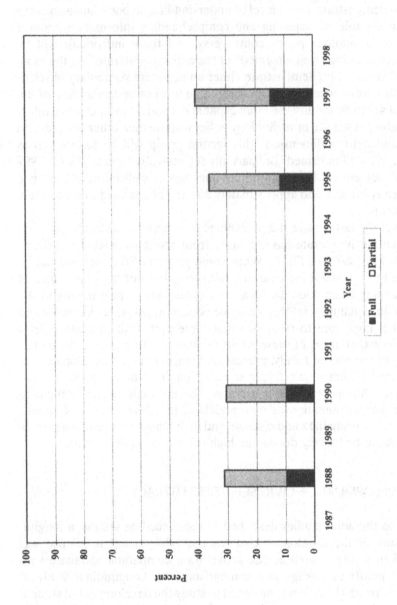

Figure 1. Civic scientific literacy among American adults, 1988–1997.

middle schools and 3,000 10th-grade students from 50 high schools were selected for the study. Each student was asked to complete an extensive personal questionnaire at the beginning and end of the school year and to take a science achievement test and a mathematics achievement test each October. Background data were collected from all science and mathematics teachers working in the 100 schools in the study, and individual course reports were collected from each science and mathematics teacher who served one or more LSAY students in a class. Each school principal was asked for school reports periodically to collect institutional measures. To provide measures of home and family influence, one parent of each participating student was annually interviewed by telephone for about 25 minutes. As a result of this process, the LSAY has built a record of approximately 7,000 variables for each student over a seven-year period, making it the most intensive study of the development of science and mathematics interest and competence ever undertaken.

As Freedman (1997) recently noted, "the search for a viable model of science instruction that will increase student achievement in science has become a global agenda." Working from the LSAY database, one could construct a set of structural equation models[1] that trace the development of student achievement in science from seventh grade through four years after high school, which would be the end of undergraduate work for those students continuing formal study. Through these models, one could identify the factors that are primarily responsible for the development of student competence in science and determine whether a high level of competence in school science is related to subsequent civic scientific literacy as adults.

It is necessary to begin with a description of the LSAY measure of student competence in science. The LSAY utilized the item pool from the National Assessment of Educational Progress (NAEP), a national study that provides baseline measures of student achievement in science, mathematics, reading, writing, citizenship, and other subjects (Mullis and Jenkins, 1988). The NAEP studies, however, are cross-sectional, not longitudinal, and the test questions were developed for that purpose. In its pilot year (1986–1987), the LSAY conducted an extensive field test of the NAEP item pool and demonstrated that it was possible, using Item–Response–Theory (IRT) techniques, to utilize these items to measure change over time (Bock

---

[1] In general terms, a structural equation model is a set of regression equations that provide the best estimate for a set of relationships among several independent variables and one or more dependent variables. For all of the structural analyses presented in this report, the program LISREL was used, which allows the simultaneous examination of structural relationships and the modeling of measurement errors. For a more comprehensive discussion of structural equation models, see Hayduk (1987) and Jöreskog and Sörbom (1993).

and Zimowski, 1997; Hambleton *et al.*, 1991). A series of science and mathematics achievement tests were developed, keeping a common core of linked items and rotating other items to make the tests grade appropriate and avoid excessive repetition in the items. Since the scores are computed in an IRT format, they can be converted to any of several scales. For this analysis, the science achievement scores of seventh-grade students were set to a mean of 50, with a standard deviation of 10. All of the scores of students in subsequent years were calibrated on this base scale.

Using this metric, researchers found the mean level of student achievement in science grew from a mean score of 51.5 in grade seven to a mean score of 65.1 in grade 12 (see Fig. 2). As one can see, the rate of achievement growth during the high school years is relatively low, and the models described in the following text will provide some explanation of this pattern. Further, an examination of the mean scores for students of the least educated parents (those who did not complete high school) and the best-educated parents (those who have a graduate or professional degree) shows that the rate of growth is more positive for students from better educated parents and almost flat for students from lesser educated parents. The cumulative impact of parent and home influence will be illustrated in the following analyses.

A Model to Predict Ninth Grade Student Science Achievement

The path model, used to predict science achievement in ninth graders, is relatively complex and difficult to interpret visually (see Fig. 3). For readers unfamiliar with structural equation models, it may be useful to describe the general form and logic of the model. In a path model, all variables to the left of the model are assumed to be either chronologically or logically prior to any variable to its right. The basic idea is that prior variables can influence subsequent variables and that absence of influence—shown by a path—means the prior variable did not influence the subsequent variable. The magnitude of any influence or association between two variables is reflected in the path coefficient, which is the standardized beta coefficient from a regression equation predicting the variable at the end of the path. As in all path analyses, it is possible to estimate the total influence of each variable on the predicted variable by multiplying the path coefficients in all possible paths leading to the predicted variable.

An examination of the estimated total effects of each of the variables used to predict grade nine science achievement may make the model more comprehensible. Rather than provide a comprehensive description of all of the variables used in the model at the outset, the variables in this and subsequent models will be defined as they are encountered in these analyses.

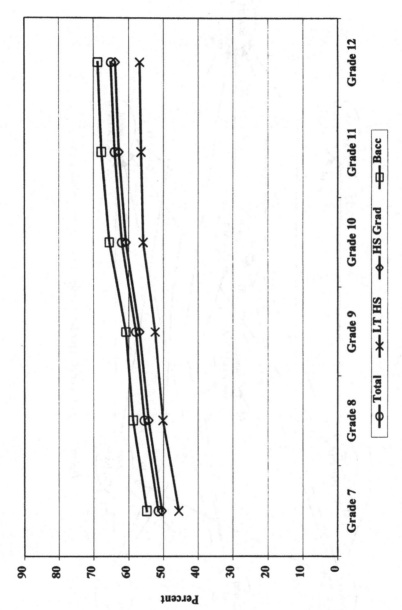

Figure 2. Science achievement scores for grades 7–12, for all students and by level of parent education.

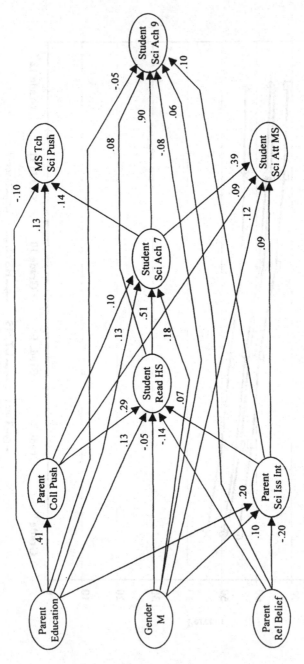

Figure 3. A path model to predict science achievement in ninth grade.

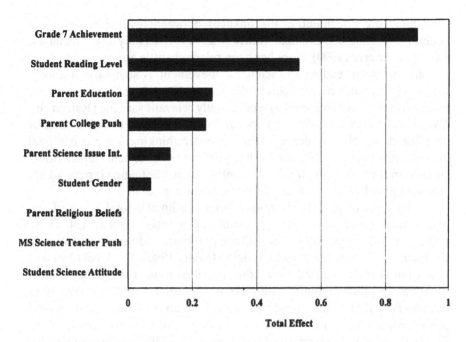

Figure 4. Estimated total effects of selected variables in the prediction of student science achievement in ninth grade.

The strongest predictor of science achievement in ninth grade was student science achievement at seventh grade, with a total effect of .90 (see Fig. 4). This result indicates that there is little change in the relative standing of students in science achievement during the middle school years. This pattern suggests that the basic ranking of students in regard to the level of science understanding occurs prior to middle school and may reflect a general stratification of students by reading and other basic academic skills during the elementary school years. These data provide no evidence of any significant catch-up during the middle school years. This pattern is consistent with the modest growth rates found in the National Assessment of Educational Progress and other national time-series studies, all of which imply a high level of stability without actually measuring individual student change (Mullis and Jenkins, 1988) and with previous analyses of longitudinal change during the middle school years (Hoffer, 1992).

The second strongest predictor of ninth grade science achievement was a composite high school reading score,[2] with a total effect of .53. Despite

---

[2] Using a set of four readings taken from the High School and Beyond study (HSB) and the National Elementary Longitudinal Study of 1988 (NELS88), the reading test included 15 items. The Index of High School Reading Ability is a composite score for reading tests taken in ninth and 12th grades. The Index's range is 0–15.

solid literature documenting the linkages between reading, writing, and general academic achievement (Paris *et al.*, 1991; Tierney and Shanahan, 1991; Langer *et al.*, 1990), there has been relatively little analysis of the relationship between reading and science achievement. A summary of important input and outcome variables by the National Research Council does not mention student reading level as a potentially relevant variable (Raizen and Jones, 1985). Resnick (1987) has advanced the argument that reading of complex material is evidence of higher-order thinking, but no National Assessment study has collected both reading and science achievement measures from the same students. In this context, then, this finding is particularly interesting and suggests the need for additional research.

The level of parent education[3] had an estimated total effect of .26, and a measure of parent college push[4] had a total effect of .24. These patterns reflect the structural advantage that better-educated parents gain for their children in the present system (Oakes, 1990). The level of parent education is often omitted from educational analyses of student achievement on the grounds that it is an existing condition outside the control of schools. Few studies have tried to separate the level of parent educational push or encouragement from the level of parent education (Keeves, 1975). The significant parallel impact of parent education and parent college push suggest these are important factors in understanding student science achievement and need to be taken into account regularly in educational research.

The mean science achievement score of children of parents with strong fundamentalist religious views[5] did not differ significantly from the scores of students whose parents' religious views were less conservative.

---

[3] Parent education is a measure of the highest level of formal education completed either by a parent in two-parent families or by the single parent in one-parent families. Previous analyses have found that the highest level of educational attainment by either parent is a better predictor of most student outcome measures than either the mean level of parent education or the education of either parent.

[4] Parent college push is a composite measure of the highest level of education that parents want their child to attain and the level of disappointment they would feel if the child failed. Parent expectations that their student would complete a graduate or professional degree was assigned a value of three; expectations of a baccalaureate were assigned a value of 2; expectations of completion of high school were assigned a value of 1. A high level of disappointment was assigned a value of 2; a moderate level of disappointment was assigned a value of 1. The two values were summed, producing an index of 0–5. An index score of 0 would reflect a child whose parents did not expect high school completion, and a score of 5 would reflect a child whose parents expect a graduate or professional degree and would be very disappointed if that child did not attain it.

[5] A measure of parent religious views was constructed, using their agreement or disagreement with three items: "There is a personal God who hears the prayers of individual men and

Cumulatively, these variables indicate parental transmission of language skills, parental encouragement of education, and parent education provide the driving force in student science achievement, and it appears that parental religious views do not significantly modify the transmission of these basic educational skills and goals.

Given the central position of science issues in the science, technology, and society movement, this analysis will include a set of variables measuring parent and student interest in science and technology issues[7] and seek to understand the influence of these variables on student science achievement. The level of parent interest in science and technology issues had a total effect of .13, indicating that the students of parents with more interest in science and technology issues were likely to score slightly higher on the ninth grade science achievement test than would children whose parents had little interest in scientific or technical issues. Following the pattern found in the literature (Rennie and Punch, 1991), student attitude toward science[6] during seventh and eighth grades was unrelated to science achievement.

Boys tended to score higher than girls (.07) on the ninth grade science achievement test, holding constant parent education, parent religious views, parent college push, and reading ability. Since the student's gender is treated as a dichotomous variable in these analyses, with males having a code of one and females having a code of zero (Code assignment does not reflect a value judgment), a positive coefficient or total effect means that boys scored higher and a negative coefficient means that girls scored higher. This pattern is consistent with three decades of literature examining gender differences in science (Ormerod and Duckworth, 1975; Young and Fraser, 1994; Catsambis, 1995).

---

women"; "The Bible is the actual word of God and is to be taken literally, word for word"; and "Human beings, as we know them, developed from earlier species of animals." Each respondent was classified as liberal–humanist, moderate–mainstream, conservative, or fundamentalist on the basis of combinations of these responses.

[6] The measure of student attitude toward science was the mean score on four subscales: interest, ability, utility, and anxiety. Individual items included in the subscales reflected student agreement or disagreement with the following items: "I enjoy science"; "I enjoy my science class"; "Math is useful in everyday problems"; "Science is useful in everyday problems"; "It is important to know science to get a good job"; "I will use science in many ways as an adult"; "I am good at science"; "I usually understand what I am doing in science"; "Doing science often makes me nervous or upset"; "I often get scared when I open my science book and see a page of problems." Using Likert scoring, the index range was 0–20.

[7] For both parents and students, science issue interest was measured by their reported level of interest (very interested, moderately interested, and not interested) in new scientific discoveries, the use of new inventions and technologies, and issues about space exploration. A report of very interested was given 2 points; a response of moderately interested was given 1 point; a report of no interest was given no points. Summed across the three issues, the index of science issue interest ranges was 0–6.

A Model to Predict 12th Grade Student Science Achievement

By the last year of high school, the strongest predictor of 12th grade science achievement was ninth grade science achievement (.84). Seventh grade science achievement was the second strongest predictor (.75). This pattern reinforces the previous observation that the relative ordering of students in regard to science achievement changed very little during the middle school or high school years (see Fig. 6) and is consistent with other analyses of longitudinal change during the high school years (Reynolds and Walberg, 1992). Reading skill was the third strongest predictor (.54), indicating that basic reading skills are important to success in science throughout the middle-school and high-school years. Consistent with the literature (Schibeci, 1989; Oakes, 1990), the combination of reading and prior achievement provide a *de facto* stratification of students that persists from at least the beginning of middle school through the end of high school.

The level of parent education was the fourth strongest predictor, with an estimated total effect of .28. Parent college push had a total effect of .26. The model shows that the children of better-educated parents have higher reading scores (.26), are placed in a higher science track by ninth grade (.12), are likely to take more laboratory science courses in high school (.38), and receive more parental encouragement to plan for college (.41) than the children of lesser educated parents.

Among high school students, parent religious views had a negative total effect of −.10 on science achievement. The children of religious fundamentalist parents tended to score lower on reading (−.16) and to express lower levels of interest in science and technology issues (−.09). The level of parent interest in science and technology issues during the middle school years continued to have a small positive influence (.12) on 12th grade student science achievement.

The major source of school influence is the number of laboratory science courses taken, with an estimated total effect of .16 (see Fig. 5). Substantial literature supports the positive contribution of laboratory experiences on student science achievement (Gunsch, 1972; Johnson *et al.*, 1974; Dickinson, 1976; Freedman, 1997).

By the end of high school, the level of middle-school and high school science teacher encouragement had no significant influence on the level of student science achievement. The level of student science track location in ninth grade is also unrelated to 12th grade science achievement, indicating that prior parental influence and subsequent high school course-taking fully account for the ninth grade track placement.

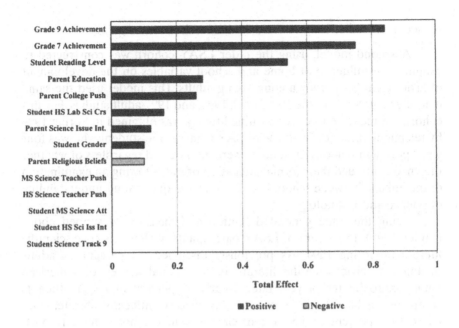

Figure 5. Estimated total effects of selected variables in the prediction of student science achievement in 12th grade.

Boys continued to score slightly higher in science achievement than girls (.10), holding constant parental influence, prior achievement, and high school course-taking patterns. While a full analysis of the differential influence of gender is beyond the scope of this analysis, it is useful to note that this total effect is a net effect, including both some positive and negative influences for each gender group. Girls, for example, have higher reading scores while boys express a higher level of interest in science issues. Gender was not related to the number of high school laboratory science courses taken.

The results from the whole model suggest that early successes and advantages multiply in the school systems of the United States and may be attributed to better-educated parents and the early development of strong reading skills. There appears to be relatively little re-ordering of students in regard to science achievement during the middle school or high school years. The pattern of mean science achievement scores suggests only modest levels of growth in science achievement during the high school years.

A Model to Predict Student Civic Scientific Literacy at Age 22

A second model, using the older LSAY cohort, was constructed to examine the influence of home and school variables on the development of civic scientific literacy among young adults. This model used the same measures employed in the 1998, 1990, 1995, and 1997 adult studies. For this cohort, the measure of civic scientific literacy was obtained in Spring 1994 by telephone interview with each LSAY Cohort One student who was four years post high school. All students were interviewed regardless of whether they had continued their formal education or not, allowing an examination of the linkage between school science and subsequent adult understanding of science and technology.

Using the same general definition and measures described above (Miller, 1998), 16 percent of LSAY participants qualified as civic scientific literate in Spring 1994. As previously discussed, 12 percent of adults qualified as civic scientific literate in a national survey of American adults approximately a year later. Nearly 30 percent of LSAY students completing a baccalaureate qualified as civic scientifically literate, compared to 4 percent of LSAY participants who did not enter any post-secondary study.

A similar measure of civic scientific literacy was collected in Spring 1990 from at least one parent of each student in a telephone interview during the student's senior year of high school. Approximately 10 percent of these LSAY parents qualified as civic scientifically literate, which was virtually identical to the percentage of adults in a 1990 national study that qualified as civic scientifically literate. These data will allow an examination of the linkage between parents and students in regard to civic scientific literacy.

Given the complexity of this model, it is useful to look primarily at the estimated total effects on the level of civic scientific literacy of a student who has reached age 22. The two strongest predictors of civic scientific literacy among young adults were the level of student science achievement in 10th grade and student high school reading level, with estimated total effects of .55 and .45 respectively (see Fig. 6). This reading and science achievement linkage confirms the importance of formal schooling. This result does not preclude improvement in civic scientific literacy during the adult years through informal education activities, but it does suggest that much of informal education may largely reinforce or enhance basic science learning during the school years.

The level of parent education was the third strongest predictor of young adult civic scientific literacy, with an estimated total effect of .37. Parent college push had an estimated total effect of .27. Parent civic

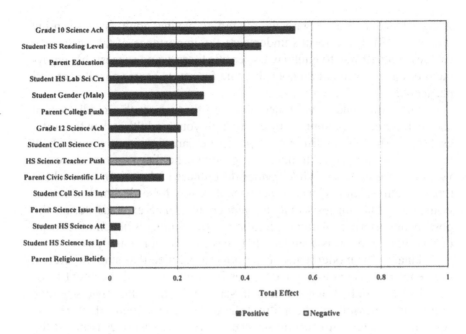

Figure 6. Estimated total effects of selected variables in the prediction of civic scientific literacy at age 22.

scientific literacy had an estimated total effect of .16. Parent interest in science and technology issues during the student's high school years displayed a small negative relationship to young adult civic scientific literacy (−.07). This may reflect the relatively stronger influence of the other parent variables. Parent religious belief was unrelated to young adult civic scientific literacy, holding constant the other variables in the model. The influence of parent education has been pervasive through these models, pointing toward a social class effect based primarily on education. Not only do better-educated parents teach their children to read earlier and more skillfully than the children of less-well-educated parents, but they provide stronger encouragement and more resources. Schools and teachers reward this advantage with more advantage. And, presumably, these better-educated young adults will do the same for their children. It is not a closed system, as in a caste system. There is a good deal of mobility within the system, but there is an undeniable structuring to the system.

The number of high school laboratory science courses taken and the number of college science courses taken were positively related to young adult civic scientific literacy, with an estimated total effects of .31 and .19, respectively (see Fig. 6). The level of encouragement by high school science teachers was negatively related to young adult civic scientific literacy (–.18). These results underscore the importance of taking science courses as a pathway to adult civic scientific literacy and suggest that enjoyment or encouragement do not substitute for actual course enrollment and participation.

The same gender difference observed throughout the middle school and high school data appears to persist into young adulthood, with a total estimated effect of .28. This means young men are significantly more likely to be civic scientific literate than young women, holding constant the other variables in the model. While gender differences in American high school science courses and science achievement scores have declined in recent years, this result suggests that the gender difference expands during the post-secondary years. A full exploration of this finding is beyond the scope of this chapter, but it is a subject that begs for additional analysis.

Finally, this model lends no support to the idea that student interest in science and technology policy issues is an effective enhancement to or incentive for student achievement in science. Young adult civic scientific literacy is a measure of the ability to read and comprehend the scientific content associated with current science and technology policy issues at the level that they might be discussed in the Tuesday *New York Times* or comparable media. It is not a measure of science knowledge comparable to the Scholastic Aptitude Test (SAT) or other tests for potential science majors. Given the citizenship orientation of this measure, it would be reasonable to expect that parental and student interest in science and technology policy issues would be strongly and positively associated with civic scientific literacy. The model, however, indicates that a high school student's interest in science and technology issues is unrelated to young adult civic scientific literacy, and the parent interest in these issues during the student's high school years and young adult interest in these issues during these initial post-secondary years displayed small negative relationships to young adult civic scientific literacy, with estimated total effects of –.07 and –.09 respectively. This result raises important questions about the tendency of many science educators to view issue-oriented courses as an important gateway to student acquisition of science understanding.

CONCLUSION

This analysis examined the level of civic scientific literacy in the United States and found a gradual increase in the percentage of American adults who qualify as civic scientifically literate and are likely to be able to read about, understand, and participate in a public policy debate involving science or technology. The analysis explored the origins of student interest and achievement in science during the middle-school and high-school years.

A set of structural equation models found a strong linkage between the learning of science during these school years and the persistence of this knowledge into the young adult years. The implications of these results are substantial. It has been assumed too often that *school science* was old at the time of delivery and had little long-term relevance, particularly in adults 20 or more years post high school. Shamos (1995) has argued this point, based largely on his own classroom observations. These results indicate that the number of science courses taken in high school and college are strong predictors of civic scientific literacy. The level of science achievement at 10th grade was the strongest predictor of civic scientific literacy among 22-year-olds. While these basic models need to be refined and expanded, this basic linkage is an important indicator that pre-college and post-secondary science courses have important life-long consequences.

Equally important, the models point to a stratification of student science achievement at early middle school years that persists throughout high school and the post-secondary years. It appears this differential is widened by enrollment in and completion of a baccalaureate program. Nearly 30 percent of young adults graduating from college qualified as civic scientifically literate—more than twice the proportion found in the total adult population. These models suggest this early stratification in science achievement is strongly related to parent and home factors. Better-educated parents have a high level of commitment to enhancing their children's life chances. Thus, they teach and try to improve their children's early reading skills. These parents provide home learning resources, such as computers and microscopes, and stress the importance of success in school. Parents who give their four-year-old child their first college tee shirt give more than a piece of clothing.

These models indicate that the early advantages provided by better-educated parents are recognized and multiplied by formal schooling. This finding does not mean efforts to enhance the life changes of disadvantaged children are hopeless, but it does mean intervention programs that ignore the substantial influence of parents and exclude them from the process are less likely to succeed.

These findings have important implications for the future of our democratic form of government. The number of important public policy controversies involving science and technology will increase substantially in the 21st century. While the processes of political and issue specialization will limit the proportion of the population that will become in any public policy dispute (Miller, 1983a; Miller *et al.*, 1997), a healthy democratic political system needs more than the 15 percent of American adults who

Science, Technology, and Society

currently qualify as civic scientifically literate to be able to participate in the resolution of public policy controversies. Further, the stratification of science achievement by level of parent education and related advantages means that science and technology policy disputes in the next generation could become intertwined with economic class differences and perhaps ethnic, racial, or religious differences. The future health of both the scientific and engineering communities and our democracy may depend, in part, on some growth in the proportion of Americans who are able to read and understand the issues in public policy disputes involving science and technology.

Finally, these results do not support the view that a strong emphasis on the study of science and technology issues is a viable pathway to an understanding of science. Four years after high school graduation, both student and parent interest in science and technology issues displayed a weak negative relationship to civic scientific literacy. Clearly, the path to civic scientific literacy runs through science classes in high school and college, and social science or issue-oriented classes do not appear to be viable substitutes. It may be that the study of science, technology, and society as a supplement to a solid core of science and mathematics courses will produce scientifically literate citizens who have an appreciation of both science and politics, but these data and this analysis point to a strong and essential linkage between formal schooling in science and mathematics and the development of civic scientific literacy.

REFERENCES

Almond, G. A. (1950). *The American People and Foreign Policy*, Harcourt, Brace & Company, New York.
Almond, G. A., and Verba, S. (1963). *The Civic Culture*, Princeton University Press, Princeton, NJ.
Almond, G. A., and Verba, S. (eds.) (1980). *The Civic Culture Revisited*, Little Brown & Company, Boston.
Bauer, M., Durant, J. R., and Evans, G. A. (1994). European public perceptions of science. *International Journal of Public Opinion Research* 6:163–186.
Bock, R. D., and Zimowski, M. F. (1997). Multiple-group IRT. In van der Linden, W. J. and Hambleton, R. K. (eds.), *Handbook of Modern Item Response Theory*, Springer-Verlag, New York.
Bosso, C. J. (1987). *Pesticides and Politics: The Life Cycle of a Public Issue*, University of Pittsburgh Press, Pittsburgh, PA.
Carson, R. (1962). *Silent Spring*, Houghton Mifflin, New York.
Catsambis, S. (1995). Gender, race, ethnicity, and science education in the middle grades. *Journal of Research in Science Teaching* 32:243–257.
Costner, H. L. (1965). Criteria for measures of association. *American Sociological Review* 30(3):341–353.

Dickinson, D. H. (1976). Community college student's achievement and attitude change in lecture only: Lecture-laboratory approach to general education biological science courses (Doctoral dissertation, Utah State University, Provo, Utah).

Durant, J. R., Evans, G. A., Thomas, G. P. (1989). The public understanding of science. *Nature* 340:11–14.

Durant, J. R., Evans, G. A., Thomas, G. P. (1992). Public understanding of science in Britain: The role of medicine in the popular presentation of science. *Public Understanding of Science* 1:161–182.

Evans, G. A., and Durant, J. R. (1995). The relationship between knowledge and attitudes in the public understanding of science in Britain. *Public Understanding of Science* 4:57–74.

Freedman, M. P. (1997). Relationship among laboratory instruction, attitude to science, and achievement in science knowledge. *Journal of Research in Science Teaching* 34:343–357.

Freudenburg, W. R., and Rosa, E. A. (1984). *Public Reactions to Nuclear Power.* Westview Press, Boulder, CO.

Goodman, L. A., and Kruskal, W. H. (1954). Measures of association for cross-classifications. *Journal of the American Statistical Association* 49:732–764.

Gunsch, L. (1972). A comparison of student's achievement and attitude changes resulting from a laboratory and non-laboratory approach to general education physical science courses (Doctoral dissertation, University of Northern Colorado, Greeley, CO).

Harman, D. (1970). Illiteracy: An overview. *Harvard Educational Review* 40:226–230.

Hambleton, R. K., Saminathan, H., and Rogers, H. J. (1991). *Fundamentals of Item Response Theory,* Sage, New York.

Hayduk, L. A. (1987). *Structural Equation Modeling with LISREL.* The Johns Hopkins University Press, Baltimore, MD.

Hoffer, T. B. (1992). Middle school ability grouping and student achievement in science and mathematics. *Educational Evaluation and Policy Analysis* 14:205–227.

Irwin, A., and Wynne, B. (eds.) (1996). *Misunderstanding science? The public reconstruction of science and technology,* Cambridge University Press, Cambridge, U.K.

Johnson, R. T., Ryan, F. L., and Schroeder, H. (1974). Inquiry and the development of positive attitudes. *Science Education* 58:51–56.

Jöreskog, K., and Sörbom, D. (1993). *LISREL 8.* Scientific Software International, Chicago.

Kaase, M., and Newton, K. (1995). *Beliefs in Government,* Oxford University Press, Oxford, UK.

Keeves, J. P. (1975). The home, the school, and achievement in mathematics and science. *Science Education* 59:439–460.

Langer, J., Applebee, A., Mullis, I., and Foertsch, M. (1990). *Learning to Read in Our Nation's Schools,* Educational Testing Service, National Assessment of Educational Progress, Princeton, NJ.

Lau, R. R., and Sears, D. O. (eds.), (1986). *Political Cognition,* Lawrence Erlbaum Associates, Hillsdale, NJ.

Milburn, M. A. (1991). *Persuasion and Politics: The Social Psychology of Public Opinion,* Brooks/Cole, Pacific Grove, CA.

Miller, J. D. (1983a). *The American People and Science Policy,* Pergamon Press, New York.

Miller, J. D. (1983b). Scientific literacy: A conceptual and empirical review. *Daedalus* 112(2):29–48.

Miller, J. D. (1987a). Scientific literacy in the United States. In Evered, D., and O'Connor, M. (eds.), *Communicating Science to the Public,* Wiley, London.

Miller, J. D. (1987b). The Challenger accident and public opinion. *Space Policy* 3(2):122–140.

Miller, J. D. (1989). Scientific literacy. A paper presented to the 1989 annual meeting of the American Association for the Advancement of Science, January 17, 1989.

Miller, J. D. (1995). Scientific literacy for effective citizenship. In Yager, R. E. (ed.), *Science/Technology/Society as Reform in Science Education*, State University Press of New York, New York.

Miller, J. D. (1998). The measurement of civic scientific literacy. *Public Understanding of Science* 7:1–21.

Miller, J. D., Pardo, R., and Niwa, F. (1997). *Public Perceptions of Science and Technology: A Comparative Study of the European Union, the United States, Japan, and Canada*, BBV Foundation, Madrid, Spain.

Minsky, M. (1986). *The Society of Mind*, Simon and Schuster, New York.

Mintz, M. (1965). *The Therapeutic Nightmare*, Houghton Mifflin, New York.

Morone, J. G., and Woodhouse, E. J. (1989). *The Demise of Nuclear Energy?*, Yale University Press, New Haven, CT.

Mullis, I. V. S., and Jenkins, L. B. (1988). *The Science Report Card: Elements of Risk and Recovery*, Educational Testing Service, Princeton, NJ.

National Science Board (1981). *Science Indicators—1980*, U.S. Government Printing Office, Washington, D.C.

National Science Board (1983). *Science Indicators—1982*, U.S. Government Printing Office, Washington, D.C.

National Science Board (1986). *Science Indicators—1985*, U.S. Government Printing Office, Washington, D.C.

National Science Board (1988). *Science and Engineering Indicators—1987*, U.S. Government Printing Office, Washington, D.C.

National Science Board (1990). *Science and Engineering Indicators—1989*, U.S. Government Printing Office, Washington, D.C.

National Science Board (1992). *Science and Engineering Indicators—1991*, U.S. Government Printing Office, Washington, D.C.

National Science Board (1994). *Science and Engineering Indicators—1993*, U.S. Government Printing Office, Washington, D.C.

National Science Board (1996). *Science and Engineering Indicators—1996*, U.S. Government Printing Office, Washington, D.C.

National Science Board (1998). *Science and Engineering Indicators—1998*, U.S. Government Printing Office, Washington, D.C.

Nelkin, D. (1977). *Technological Decisions and Democracy: European Experiments in Public Participation*, Sage, Beverly Hills, CA.

Noll, V. H. (1935). Measuring the scientific attitude. *Journal of Abnormal and Social Psychology* 30:145–154.

Oakes, J. (1990). *Multiplying Inequalities: The Effects of Race, Social Class, and Tracking on Opportunities to Learn Mathematics and Science*, The RAND Corporation, Santa Monica, CA.

Ormerod, M. B., and Duckworth, D. (1975). *Pupils' Attitudes to Science*, NFER Publishing Company, Berks, UK.

Paris, S. G., Wasik, B. A., and Turner, J. C. (1991). The development of strategic readers. In Barr, R., Kamil, M., Mosenthal, P., and Pearson, P. D. (eds.), *The Handbook of Reading Research: Volume II*, Longman, New York.

Pick, H. L., van den Broek, P., and Knill, D.C. (eds.), (1992). *Cognition: Conceptual and Methodological Issues*, American Psychological Association, Washington, D.C.

Popper, K. R. (1959). *The Logic of Scientific Discovery*, Basic Books, New York.

Popkin, S. L. (1994). *The Reasoning Voter*, University of Chicago Press, Chicago.

Raizen, S. and Jones, L. V. (1985). *Indicators of Precollege Education in Science and Mathematics: A Preliminary Review*, National Academy Press, Washington, D.C.

Rennie, L. J., and Punch, K. F. (1991). The relationship between affect and achievement in science. *Journal of Research in Science Teaching* 28:193–209.

Resnick, L. B. (1987). *Education and Learning to Think*, National Academy Press, Washington, D.C.

Resnick, D. P., and Resnick, L. B. (1977). The nature of literacy: An historical exploration. *Harvard Educational Review* 47:370–385.

Ressmeyer, T. J. (1994). Attentiveness and mobilization for science policy (Doctoral dissertation, Northern Illinois University, DeKalb, IL).

Reynolds, A. J., and Walberg, H. J. (1992). A structural model of science achievement and attitude: An extension to high school. *Journal of Educational Psychology* 84:371–382.

Rosenau, J. A. (1974). *Citizenship Between Elections*, Free Press, New York.

Schank, R. (1977). *Scripts, Plans, Goals, and Understanding*, Lawrence Erlbaum Associates, Hillsdale, NJ.

Schebeci, R. A. (1989). Home, school, and peer group influences of student attitudes and achievement in science. *Science Education* 73:13–24.

Shamos, M. (1995). *The Myth of Scientific Literacy*, Rutgers University Press, New Brunswick, NJ.

Shen, B. S. J. (1975). Scientific literacy and the public understanding of science. In Day, S. (ed.), *Communication of Scientific Information*, Karger, Basel, Switzerland.

Tierney, R., and Shanahan, T. (1991). In Barr, R., Kamil, M., Mosenthal, P., and Pearson, P. D. (eds.), *The Handbook of Reading Research: Volume II*, Longman, New York.

Verba, S., Nie, N. H., and Kim, J. (1978). *Participation and Political Equality: A Seven-nation Comparison*, Cambridge University Press, Cambridge, UK.

Verba, S., Schlozman, K. L., and Brady, H. E. (1995). *Voice and Equality: Civic Voluntarism in American Politics*, Harvard University Press, Cambridge, MA.

Wynne, B. (1991). Knowledges in context. *Science, Technology & Human Values* 16:111–121.

Young, D. J., and Fraser, B. J. (1994). Gender differences in science achievement: Do schools make a difference? *Journal of Research in Science Teaching* 31:857–871.

Ziman, J. (1991). Public understanding of science. *Science, Technology & Human Values* 16: 99–105

CHAPTER 3

STS Science in Canada
From Policy to Student Evaluation

Glen S. Aikenhead

This chapter will map out the territory educators should explore if they expect to develop STS science curricula. How are successful STS curricula produced? In Canada, we have had about 25 years experience with research and development in STS science teaching. In 1997 this experience culminated in a national STS science framework.

The territory to be mapped out in this chapter includes a range of four tasks. Each task involves a process that leads to a desired product. The relationships among four fundamental processes and products are summarized in Table 1. Table 1 relates the *processes* of deliberation—research and development (R&D), implementation, and instruction–assessment—with their associated *products*: curriculum policy, classroom materials, teacher understanding, and student learning, respectively. The sequence across Table 1 (from policy to student learning) reflects three levels of curriculum: 1) intended curriculum (government policy), 2) translated curriculum (textbooks and teachers' ideas about what will be taught), and 3) learned curriculum (the concepts, capabilities, and attitudes that students actually take

Glen S. Aikenhead, Department of Curriculum Studies, College of Education, University of Saskatchewan, Saskatoon, SK S7N 0X1, Canada

Science, Technology, and Society: A Sourcebook on Research and Practice, edited by Kumar and Chubin, Kluwer Academic / Plenum Publishers, New York, 2000.

Table 1. Relationships between Processes and Products in STS Science Education

	PRODUCTS			
	Curriculum Policy	Classroom Materials	Teacher Understanding	Student Learning
PROCESSES				
Deliberation	**high**	low		
Research and Development	low	**high**	low	
Implementation		low	**high**	low
Instruction-Assessment			low	**high**

away with them). STS science educators must consider all three levels of the curriculum before successful STS courses can be produced.

This chapter is organized around the four product–process pairs designated in Table 1 as having a "high" relationship.

An alternative approach to the one taken in this chapter was proposed by Cheek (1992). His "STS Curriculum Development Model" included features such as: theoretical considerations (constructivism, moral development, critical thinking, etc.), data collection considerations (children's views about the scientific, technological and social aspects of the curriculum content), content considerations (technology content, values, skills, etc), delivery system considerations (teacher knowledge, student readiness, etc.), curriculum materials design, implementation, and evaluation.

Space does not permit a detailed discussion of every process and product listed in Table 1. Thus, this chapter will only do a cursory exploration of the territory. Emphasis will be given to the first product–process pair: curriculum policy and deliberation. Discussions will reflect our experiences with STS science education in Canada, though these experiences are certainly related to international developments in STS, and so some of these will be noted as well.

SCIENCE EDUCATION IN CANADA

Those unfamiliar with Canadian culture need to know that education is a fiercely guarded provincial responsibility. As a result, we have a different educational system in each province. Up until now, provinces have independently developed their own science curricula, and they have even arranged for the publication of accompanying textbooks. Although this provincial independence has been a source of pride to provincial educators, it has been an expensive source of pride because of the duplication of effort and unnecessary disparity among the provinces.

Canada inaugurated its first national framework for science education in October 1997, the *Common Framework of Science Learning Outcomes* developed by the Council of Ministers of Education of Canada (CMEC, 1997). This pan-Canadian protocol for collaboration on school curriculum established a science-technology-society-environment (STSE) approach to achieving scientific literacy in Canada. The publication is a curriculum policy document and not a curriculum per se. It represents a type of STS science education that may be attractive to STS educators elsewhere. The pan-Canadian protocol attempted to balance the two historical factors of provincial independence versus duplication and unnecessary disparity among the provinces.

In keeping with Canadian culture, the *Common Framework of Science Learning Outcomes* (the *Framework*) evolved through negotiation and compromise among provincial bureaucrats, who were advised by interested parties (stakeholders) in each province. This political process, however, did not meet the standards of curriculum policy development held by the Canadian science education academic community. This problem was aired at two national symposia held during annual meetings of the Canadian Society for the Study of Education in June 1997 and May 1998.

The academic science educator's discontent arose from the dramatic difference between the standards guiding the CMEC's political process and the standards used earlier by a national education policy project, funded and directed by the Science Council of Canada (SCC, 1984; Orpwood, 1985). The SCC project had painstakingly conducted its education study with the highest of scholarly standards. (This reform project will be described when the process of deliberation is explored.) These science educators expected the SCC's high standards of excellence in formulating curriculum policy would guide future policy discussions in Canada. The *contrast* between the processes followed by the two agencies (the Council of Ministers of Education of Canada [CMEC] versus the SCC) led to most of the discontent felt by university science educators. Certainly the SCC science education study's conclusions in the early 1980s did influence the CMEC's bureaucratic negotiations and compromises held in the mid 1990s.

In spite of the discontent over the process that produced the CMEC's *Framework*, the *Framework* will likely be as influential in Canada as the National Research Council's *Standards* are in the United States (NRC, 1996). Canada now has an STSE framework for science curricula across the country.

CURRICULUM POLICY AND DELIBERATION

Curriculum policy and deliberation is the first of the four pairs of products–processes shown in Table 1. Fensham (1992) pragmatically described the politics of curriculum policy as societal interest groups (stakeholders) competing for privilege and power over the curriculum. For example, school science (quite often physics) can be used to screen out students belonging to marginalized social groups (some minorities within a country, for example), thereby providing high status and social power to the more privileged students who make it through the science "pipeline" and enter science-related professions. Fensham categorized this societal self-interest as *political*. His other categories were: *economic* interests of business, industry, and labor, for a skilled work force; university scientists' interests in *maintaining their discipline*; societal groups' interests for empowerment in a nation whose *culture and social life* are influenced by science and technology; and students' interests for *individual* growth and satisfaction. As Fensham (1992) warned, the science curriculum is a social instrument that serves the interests of those who have a stake in its function and content. Therefore, stakeholders must be involved in reforming curriculum policy. The most effective way to involve them is through deliberation.

When we consider curriculum policy by itself, several aspects must be explored:

1. *Function*—What are the goals and objectives for teaching the content?
2. *Content*—What is worth learning?
3. *Structure*—How should the science and STS content be integrated and contextualized?
4. *Sequence*—How should the teaching be organized?
5. *The process of establishing the function, content, structure, and sequence*—Who should be involved? How should curriculum policy decisions be decided?

Each country and community must answer these questions for itself. STS education responds to the idiosyncratic needs of each educational jurisdiction (Solomon and Aikenhead, 1994). Although I offer no universal conclusions to these four aspects of STS curriculum policy, I do sketch the territory that must be addressed. Following the sections on function, content, structure, and sequence, I discuss the process of deliberation by which people can establish a curriculum policy.

Function

The functions (or goals) of STS instruction in schools have been the focus of a literature rationalizing the STS education movement. This literature is found in three different fields: science education, technology education, and social studies education. An overview of all three fields was published in *Theory into Practice* (Gilliom, Helgeson, and Zuga, 1991, 1992). *Science* education offers only one orientation toward STS education, but it is the usual orientation of the Canadian experience with STS.

What are the goals and objectives for teaching STS science? Its rationalization is documented worldwide (Aikenhead, 1980, 1985, 1994c; Bingle and Gaskell, 1994; Bybee, 1985; Cheek, 1992; Cross and Price, 1992; Eijkelhof and Kortland, 1987; Fensham, 1988; Gaskell, 1982, 1992; Hansen and Olson, 1996; Hunt, 1988; Hurd, 1975, 1989; Keeves and Aikenhead, 1995; Nagasu and Kumano, 1997; Pedretti and Hodson, 1995; Sjøberg, 1996; Solomon, 1981, 1994; Waks, 1987; Yager, 1992a, 1992b, 1996; Ziman, 1980). The earlier publications and those by Canadian scholars particularly influenced the CMEC's *Framework*.

STS science is essentially about students making sense out of their everyday life, today and for the future. But to what purpose? What are the goals of STS science education in Canada? Themes emerged from the literature listed above. These were compiled by Aikenhead (1994d) and are summarized here.

The inadequacies of traditional science education define one over-all theme. STS science is expected to reverse the existing negative trends in enrollments, achievement, and career choices. Specifically, STS science is expected to increase general interest in and public understanding of science, particularly for the bright creative students who are discouraged by what they perceive to be a boring and irrelevant curriculum (Oxford University, 1989; SCC, 1984).

STS science is also expected to fill a critical void in the traditional curriculum—the social responsibility in collective decisionmaking on issues related to science and technology (Aikenhead, 1985; Bingle and Gaskell, 1994; Gaskell, 1982). Such issues require a harmonious mix of a scientific-technical elite with an informed attentive citizenry. Together both groups will need to make complex decisions that involve "the application of scientific knowledge, technological expertise, social understanding, and humane compassion" (Kranzberg, 1991, p. 238). The pervasive goal of social responsibility in collective decisionmaking leads to numerous related goals: individual empowerment; intellectual capabilities such as critical thinking, logical reasoning, creative problem solving, and decision making; national and global citizenship, usually "democracy" or "stewardship"; socially responsible action by individuals; and an adroit work force for business and industry. These goals emphasize an induction into a world increasingly shaped by science and technology, more than they support an induction into a scientific discipline. In Canada in the early 1980s, STS curriculum policy was influenced by a series of position papers commissioned by the SCC for its education policy study (SCC, 1984). One position paper specifically addressed the pervasive goal of social responsibility in collective decision-making and students' induction into a world increasingly shaped by science and technology (Aikenhead, 1980).

Most viewpoints concerning the function or intentions of any science instruction can be described as "curriculum emphases" (Roberts, 1982, 1988). Roberts argued that science instruction *in general* has purposes defined by the answers to a student's plaintive cry: "Why are we learning this stuff, anyway?" Based on science curricula and textbooks published this century, Roberts classified the answers into seven categories:

1. *Solid foundations*—To prepare students for the next level of science courses
2. *Correct explanations*—To learn the truths of scientific knowledge
3. *Scientific skill development*—To learn the conceptual and manipulative skills required for participation in scientific inquiry
4. *Structure of science*—To learn how the academic side of science functions as an intellectual enterprise and to see the conceptual harmony that a scientist sees
5. *Self as explainer*—To help students in their personal efforts to explain natural phenomena and to make personal sense out of the nature of scientific explanations
6. *Everyday coping*—To help students understand important objects and events in their everyday lives
7. *Science, technology, and decisions*—To become aware of science in a social and technological context

The last two emphases suggest an STS approach to science instruction.

Fensham (1993) rationalized adding three more emphases to this list. One, "science in application," acknowledges what SATIS (Hunt, 1988) and many textbooks do when they add common sense applications of the science content, but do not allow the applications to determine the content or sequence of the science instruction. "Science for nurturing" was the second emphasis Fensham proposed. It stems from the ideology embraced by the environmental and the women's movements that emphasize nurturing the earth or the social needs of its inhabitants, respectively. A third emphasis was "science through making," in which students learn science through the process of making a technologically useful artifact or through solving a technological problem. An example of this in Canada is a nationally organized activity called "Science (sic) Olympics"—all the activities are actually technological in nature and the embedded science content is not part of the competition. Thus, the name "Technology Olympics" would be more accurate.

Ogawa (1995) expanded the conventional view that there is only *one* science to be recognized. He proposed a broader multicultural perspective when he argued that several legitimate sciences exist, including a community's common sense knowledge of nature. Ogawa proposed a "multiscience perspective" curriculum emphasis when Western science is taught to non-Western students. The Canadian *Framework* defines science in such a way that includes Fensham's application, nurturing, and making emphases and Ogawa's multiscience emphasis, along with Roberts' original seven emphases.

Science instruction in any classroom is carried out, consciously or not, with various combinations of the 11 emphases. A typical STS science course would likely include the "solid foundations" emphasis but may likely give it lower priority than the "science, technology, and decisions" or the "science through making" emphases. Emphases should be identified as functions of a curriculum, and this identification should be very much a part of curriculum policy. STS curriculum policy must clarify the mix of emphases that is intended for a given STS course.

Although STS science courses differ widely because of their different emphases or goals, this variation reflects differences in the *balance* among *similar* goals. In other words, most STS science courses harbor similar goals but give different priorities to those goals. This idea of balance is captured by the slogan "scientific literacy" (Hart, 1989; Roberts, 1983). As a slogan, scientific literacy provides an element of persuasion in rationalizing science programs (who can be against scientific literacy?). But the term can be useful in defining a goal cluster for a curriculum policy statement. This was the case in Canada, as elsewhere.

The *Common Framework of Science Learning Outcomes* (CMEC, 1997) defines the function of science education primarily in "A Vision for Scientific Literacy in Canada" (p. 4) along with a rationale entitled, "The Scientific Literacy Needs of Canadian Students and Society" (p. 5). The vision statement is repeated here.

> The framework is guided by the vision that all Canadian students, regardless of gender or cultural background, will have an opportunity to develop scientific literacy. Scientific literacy is an evolving combination of the science-related attitudes, skills, and knowledge students need to develop inquiry, problem-solving, and decision-making abilities, to become lifelong learners, and to maintain a sense of wonder about the world around them. Diverse learning experiences based on the framework will provide students with many opportunities to explore, analyse, evaluate, synthesize, appreciate, and understand the interrelationships among science, technology, society, and the environment that will affect their personal lives, their careers, and their future. (CMEC, 1997, p. 4)

The scientific literacy needs of Canadian students and of Canadian society are stipulated in the *Framework*'s goals that articulate its vision statement. These goals lead to "foundational statements," around which the whole framework is organized (described later).

Unlike an STS curriculum, the function of a conventional science curriculum was to prepare students for the next level of education, to teach correct answers, and to enculturate students into physics, chemistry, or biology (Aikenhead, 1996; Roberts, 1988). These functions are not ignored in STS science, but they are not given as strong an emphasis. As a result, an STS science curriculum addresses the needs of two groups of students: (1) future scientists and engineers (that is, the elite), and (2) citizens who need intellectual empowerment to participate thoughtfully in their society (that is, the attentive public or "science for all"). The inclusion of both groups of students in STS science responds to Fensham's (1992) warning that we must take into account the stakeholders competing for privilege and power over the science curriculum.

Content

There is a marked difference between the content of university STS courses and the content of high school STS science courses. University courses invariably deal with science and technology policy, development, and discourse (Layton, 1994; McGinn, 1991). The subject matter is abstract. High school STS courses position themselves among the concrete experiences of students. These courses provide a simplified, although intellectually honest, perspective on the human and social aspects of science.

The content of STS science includes *both* science content and STS content. Here I focus on clarifying what STS content can be. In the section that follows, "Integrative Structure," I explore how this STS content can be integrated with science content.

Hansen and Olson (1996) and Bingle and Gaskell (1994) point out that many educators narrowly conceive of STS science content as dealing with social issues that connect science with a societal problem. Rosenthal (1989), and Ziman (1984) remind us, however, that there are two types of social issues in STS science:

1. Social issues *external* to the scientific community ("science and society" topics, for example, energy conservation, population growth, or pollution)
2. Social aspects of science—issues *internal* to the scientific community (the sociology, epistemology, and history of science, for example, the cold fusion controversy, the nature of scientific theories, or how the concept of gravity was invented)

The relative importance of various social issues *external* to science was a topic researched by Piel (1981) in Project Synthesis and by Bybee and Mau (1986) in their survey of international science educators.

However, STS science must invariably address the sociology, epistemology, and history of science, that is, social issues *internal* to science. The social issues internal to science have been delineated by Snow (1987). His "system of science" has three dimensions: cognitive, personal, and sociological. The *cognitive* dimension includes experimental knowledge, hypotheses, theories, laws, and empirical observations, as well as the values that underlie them (accuracy, coherence, fruitfulness, and parsimony). The *personal* dimension encompasses a scientist's social values that influence his or her research programs and nonempirical arguments. The *sociological* dimension incorporates community values, invisible colleges, credibility, journal publications, and other aspects of importance to the scientific community. In his book *An Introduction to Science Studies*, Ziman (1984) provides a thorough and systematic treatment of STS content, both external and internal to the scientific community. These ideas are reflected in the content of Canada's *Framework* (CMEC, 1997). The *Framework*'s STSE emphasis explicitly includes social issues both external and internal to science.

The international science education community holds a variety of views concerning STS content. Nevertheless, a succinct definition of STS content was offered by Aikenhead (1994d). The definition attempted to encompass the full range of views held by science educators everywhere.

STS content is comprised of an interaction between science and technology or between science and society and any one or combination of the following:

- A technological artifact, process, or expertise
- The interactions between technology and society
- A societal issue related to science or technology
- Social science content that sheds light on a societal issue related to science and technology
- A philosophical, historical, or social issue *within* the scientific or technological community. (Aikenhead, 1994d, pp. 52–53)

Diverse examples of STS content in several countries (Australia, Canada, India, Netherlands, Nigeria, United Kingdom, and United States) are found in Solomon and Aikenhead (1994). Canada's *Framework* is flexible enough to embrace the different goals and content found among the various provinces.

Integrative Structure

STS science curricula integrate various types of content in the following way:

$$STS\ science = science\ content + STS\ content$$
$$= science\ content + internal\ social\ issues$$
$$+ external\ social\ issues.$$

But how much science content is integrated with STS content? How is this integration accomplished? To answer this question (based on a number of commercial STS materials available world wide), Aikenhead (1994d) devised "Categories of STS Science," a descriptive scheme with eight categories that characterize STS science in terms of:

- Content structure—The proportion of STS content compared with traditional science content, and the way the two are combined.
- Student evaluation—The relative emphasis given to STS versus traditional content. The description is an approximate indicator of relative emphasis, rather than a prescription for classroom practice.
- Concrete examples of STS science. (Aikenhead, 1994d, p. 53)

A spectrum underlies the proposed scheme and expresses the relative importance afforded STS content in a science course. At one end of the spectrum (category one), STS content is given lowest priority compared with traditional science content, while at the other end (category eight), it is given highest priority. The eight categories of the spectrum are:

1. Motivation by STS content
2. Casual infusion of STS content
3. Purposeful infusion of STS content
4. Singular discipline through STS content
5. Science through STS content
6. Science along with STS content
7. Infusion of science into STS content
8. STS content

One can think of each category as a conveniently identified point along the spectrum. Although no particular category can be said to represent "true" STS science instruction, categories three to six do represent views most often cited by STS science leaders. Aikenhead's eight-point scheme was inspired by a similar table about technology education in an article by Fensham (1988). The eight "Categories of STS Science" are summarized here.

1. Motivation by STS Content. STS content is just mentioned by a teacher to make a lesson more interesting to students. Students are *not* assessed on the STS content. The low status given to STS content explains why this category is not normally taken seriously as STS instruction.

2. Casual Infusion of STS Content. A short study (30 minutes to two hours in length) of STS content is attached to the science topic of traditional school science, as defined by Fensham's (1993) curriculum emphasis "science in application" and exemplified by the SATIS materials (Hunt, 1988). The STS content is *not* chosen to convey cohesive themes about the social issues internal or external to science. Rather, topics are added when teaching materials are available. Students are assessed mostly on pure science content, and usually only superficially (such as memory work) on the STS content. The relative weighting of this assessment might be, for instance, 5 percent STS content and 95 percent science content.

3. Purposeful Infusion of STS Content. A series of short studies (30 minutes to two hours in length) of STS content are integrated into science topics in a traditional science course, to systematically explore the STS content. This STS content forms *cohesive* themes. Harvard Project Physics (Holton, Rutherford and Watson, 1970) is a familiar example. Students are assessed to some degree on their understanding of those STS themes, for instance, 10 percent on STS content and 90 percent on science content.

4. Singular Discipline Through STS Content. STS science courses take on a radically different look in this and subsequent categories. Instead of following the conventional content and sequence found in traditional science textbooks (categories one to three above), science content and its sequence are chosen and organized largely by the STS content. First, a curriculum policy designates what STS content will be included in a science curriculum. Then the science content is selected on a *need-to-know basis* guided by the STS content, but selected primarily from one science discipline. There will be an STS biology, an STS chemistry, and an STS physics curriculum. The American Chemical Society's (1988) *ChemCom* is a typical category four course. A listing of pure science topics in such a course could look quite similar to a listing from a category three science course, though the sequence would be very different. However, curriculum developers taking the *need-to-know* criterion very seriously might include science and technology content not found in conventional science courses, for example, Eijkelhof's (1994) STS module *Ionizing Radiation* includes the concept of dosage. In category four courses, students are assessed on their in-depth understanding of the STS content, but not nearly as extensively as they are on the pure science content, for instance, 20 percent STS content and 80 percent science content.

    5. *Science Through STS Content.* As in category four courses, STS content serves as an organizer for the science content and its sequence. But in category five courses, science content is multidisciplinary, as dictated by the STS content on a need-to-know basis. A listing of pure science topics might look like a selection of important science topics from a variety of traditional school science courses. Again, one can find science and technology content not found in conventional science courses. *Logical Reasoning in Science and Technology* (Aikenhead, 1991) in Canada, and the Science Education for Public Understanding courses (Thier and Nagle, 1994) in the United States, exemplify the inclusion of science and technology content not normally found in traditional science courses but highly relevant to an everyday event or issue. In category five courses, students are assessed on their in-depth understanding of the STS content, but not quite as extensively as they are on the pure science content, for instance, 30 percent STS content and 70 percent science.

    6. *Science along with STS Content.* STS content is the focus of instruction. Relevant science content enriches this learning. In Canada, the British Columbia Ministry of Education developed Science and Technology 11 in 1985 (Gaskell, 1989). Students are assessed equally on the STS content and pure science content.

    7. *Infusion of Science into STS Content.* STS content is a greater focus of instruction. Relevant science content is mentioned, but not systematically taught. Emphasis may be given to broad scientific principles. Materials classified as category seven could be infused into a standard school science course, yielding a category three STS science course. In Canada, the possibility of such a course existing was illustrated in Science: A Way of Knowing (Aikenhead, 1979). In a category seven course, students are primarily assessed on the STS content, and only partially on pure science content, for instance, 80 percent on STS content and 20 percent on science content.

    8. *STS Content.* A major technology or social issue is studied. Science content is mentioned but only to indicate an existing link to science. The materials classified as category eight could be infused into a standard school science course, yielding a category three STS science course. Students are *not* assessed on pure science content to any appreciable degree.

This eight-category scheme does not attempt to evaluate different approaches to STS science instruction. Nor does it attempt to prescribe any particular set of goals or goal priorities. Moreover, the scheme does not address some highly relevant issues, such as teaching methods (for example, inquiry, problem solving, decisionmaking), contexts for instruction (for example, scientific controversies, local issues, public policies, global problems), and assumptions about how students learn (though constructivism predominates in STS science [Cheek, 1992]).

Canada's *Framework* (CMEC, 1997) does not specify a particular integrative structure for its STSE science curriculum. However, when describing various grade levels, it explicitly integrates four areas of content goals (STSE, skills, knowledge, and attitudes), and science content is unambiguously contextualized within suggested STSE content. Thus, categories three to five (previously described) seem to be favored. The provinces are free to devise their own STSE content, structure, and sequence, but the *Framework*'s clear expectation is that STSE content will serve as a context for canonical science subject matter on a need-to-know basis.

The eight categories of STS science provide a language for discussing various structures for STS curricula, classroom materials, and teachers' instruction. For instance, we can expect that many science teachers will at first be more comfortable teaching a category three curriculum than a category four curriculum.

Curriculum policy should specify the category or categories of STS science that are intended in an STS science curriculum. Different parts of the same curriculum may have different structures, of course. Thus, different units within a curriculum may be characterized by different categories.

Sequence

The "Categories of STS Science" represent a general integrative structure for STS science. A particular *sequence* to follow by teachers and curriculum writers was empirically discovered by Eijkelhof and Kortland (1987, 1988). Their research and development (R&D) took place in the Dutch project *PLON* (an acronym for physics in a social context), a category four curriculum consisting of about 35 modules for grades 7–12. Eijkelhof and Kortland investigated different sequences and discovered one pattern that nurtured successful learning by students: Begin with a societal need or issue which invariably leads to a technology, which in turn creates the need to know science content, which then leads to further investigations of related technologies that finally inform a deeper understanding of the original societal need or issue. This pattern was discussed and illustrated for North American STS science courses by Aikenhead (1992a, 1994d) and is summarized here.

When an STS science unit or lesson begins, students consider a *social issue* or an *everyday event* (for example, a court case on drunken driving, or the lighting requirements of various rooms in houses, schools, and businesses). Then students become acquainted with relevant *technology* (for example, the Borkenstein breathalyzer, or architecture designs and commercial lighting fixtures). The social issue or everyday event, along with the related technology, create in the students' minds the need to know the canonical *science* that helps students make sense out of the issue or event and the technology. For example, in the case of the issue of drunken driving and the breathalyzer, students need to know mixtures, redox reactions, electrical circuits, body systems, and photometry; while in the case of light sources, students need to know photometry, eye physiology, the nature of light, and the electromagnetic spectrum. Armed with this relevant science content, students next reexamine the original technology or explore more sophisticated technology, and then move on to reexamine the original social issue or everyday event. This last step often involves making a relevant decision on the issue or event, for example, should Hoffman LaRoche develop a "sobering up" pill, a new technology? Or, what type of electrical bulbs should our home purchase? Students will make thoughtful decisions informed by an indepth understanding of the underlying science, informed by a grasp of the relevant technology, and informed by an awareness of the guiding social values inherent in various decision choices (Aikenhead, 1980, 1985).

In summary, Eijkelhof and Kortland (1987, 1988) devised the sequence:

social content → technology content → science content → advanced technology → advanced social content

Although teachers may spend the majority of their instruction time on the canonical science content (for instance, 70–80 percent of instruction time), the Eijkelhof and Kortland sequence ensures that the science content will be contextualized in a meaningful way for students. This contextualized learning is promoted in the Canadian *Framework* (CMEC, 1997).

The Process of Deliberation

Curriculum policy can be established in a number of ways. Educational jurisdictions vary in the decisionmaking processes they employ. In Canadian culture, the process of *deliberation* has shown greatest potential for success. This is a combination of "top-down" and "grass-roots" methods of policymaking. Deliberation is a structured and informed dialogue among various stakeholders. In the SCC's science education study (SCC, 1984), for example, a wide array of stakeholders were involved: science teachers, university professors, students, community leaders, parents, and government officials. An informed decision over curriculum policy was reached, based on the values held by the stakeholders and their reading of relevant research.

One major purpose of deliberation is to involve the science teachers who will eventually implement the new curriculum, and at the same time, to involve the people who can offer those teachers support, encouragement, and guidance. Roberts (1988) illustrated the need for this support by way of a case study of STS policymaking in the province of Alberta.

Inspired by Schwab's (1974) "deliberative enquiry," the SCC in the early 1980s employed the deliberation process during a large national education study (Orpwood, 1985). In its report, *Science for Every Student* (SCC, 1984), the SCC called for a renewal (reform) of science education, advising educators to teach scientific concepts and skills embedded in social and technological contexts relevant to all students. It reached this curriculum policy conclusion through the processes of research and deliberation, which occurred in three phases:

1. Issue identification—What are the problems?
2. Data collection—What are the facts?
3. Option development—Where do we go from here?

The SCC's education study ensured that significant problems were identified, that appropriate data were collected, and that these problems and data were considered by the diverse stakeholders attending one of the 11, two-day deliberative conferences held across Canada in 1983. As previously mentioned, stakeholders included high school students (science-prone and science-shy), teachers (elementary and secondary), parents, elected school officials, the scientific community, business, industry, and the labor movement, and university science educators. Although consensus was not reached at any of those deliberative conferences, a full range of interests and viewpoints were aired. In one conference, for instance, it was instructive to watch a rural elementary teacher successfully challenge the rhetoric of a corporate president representing a biotechnology firm. The deliberative conferences unfolded as Schwab (1978) had predicted:

> Deliberation is complex and arduous.... It must try to identify the desiderata in the case. It must generate alternative solutions.... it must then weigh alternatives and their costs and consequences against one another, and choose, not the *right* alternative, for there is no such thing, but the *best* one. (Schwab, 1978, pp. 318–319)

The "best" solution (a curriculum policy) published by the SCC (SCC, 1984) is summarized by a set of recommendations that included the following points (organized according to their future influence on science curriculum policy development in Canada):

1. Along with scientific concepts and skills, students should learn an appreciation for:
   a. authentic science—the nature of science and scientists, including the way science generates and uses its knowledge
   b. technology in Canada
   c. the interrelationships among science, technology, and society
2. Females should be particularly encouraged to pursue science and technology in school.
3. Academically talented students should be challenged to reason critically and creatively.
4. Student evaluation should concentrate on fundamental understandings and should reflect the complete range of goals of science teaching, rather than focus strictly on the memorization of facts and the rote application of formulas.

These recommendations, as well as the process of deliberation itself, greatly influenced a study that was designed to change the science curriculum in Saskatchewan.

The Saskatchewan science study (conducted 1986–1987) is described in detail by Hart (1989), but a few salient features will be mentioned here. Drawing in part upon the SCC's study, Hart formulated "a set of recommendations designed to illuminate discrepancies between actual and desired states of school science" (p. 610). This platform for renewal of science education in Saskatchewan was discussed by 337 educators during 18 one-day deliberative conferences held across Saskatchewan. The sponsor of Hart's study limited stakeholders to science teachers, administrators, and science consultants. Many science teachers not attending the conferences submitted to Hart written responses to the agenda items considered during the deliberative conferences.

An overwhelming 92 percent of Saskatchewan science teachers endorsed teaching science in a way that balanced seven set of goals, a goal cluster referred to as "Seven Dimensions of Scientific Literacy": (1) the nature of science; (2) the key facts, principles, and concepts of science; (3) the intellectual processes used when doing science; (4) the interactions among science, technology, and society; (5) the values that underlie science; (6) the know-how to use instruments required for doing science; and (7) personal interests and attitudes toward scientific and technological matters. Although 88 percent of the teachers believed that a STSE emphasis should be adopted, teachers expressed many concerns about: (1) the balance between the STSE emphasis and other emphases (for example, a solid foundation for the next level of science study); (2) the evaluation of students with respect to the STSE goals; (3) the availability of appropriate teaching materials; (4) the possible erosion of traditional science subject matter; and (5) the need to teach controversial issues. In other words, teachers were positive but certainly cautious about changing their science curriculum toward an STSE approach. The provincial study had assumed that most science teachers would gain a degree of ownership in a new curriculum developed on the basis of their deliberative conferences.

The Saskatchewan study illustrates one version of the process of deliberation that establishes an STS curriculum policy. Roberts (1988) described what happens when the process of deliberation is not taken seriously, as was the case in the late 1980s with the Alberta Ministry of Education's approach to revising its science curriculum into STS science. Teacher committees and bureaucrats worked conscientiously to develop a state-of-the-art STS curriculum, but resistance from the vested interests of a social elite (spearheaded by medical doctors and university science professors) ensured sufficient political intervention to stop the implementation. Blades (1997) provides an intriguing postmodern analysis of the tensions and dynamics among the principal protagonists. Roberts (1988) claimed that the main question to be resolved is "What counts as science education?":

> So the sticky question "What counts as science education?" has three characteristics. First, the answer to it requires that choices be made—choices among science topics and among curriculum emphases. Second, the answer is a defensible decision rather than a theoretically determined solution to a problem theoretically posed. Third, the answer is not arrived at by research (alone), nor with universal applicability; it is arrived at by the process of deliberation, and the answer is uniquely tailored to individual situations. Hence the answer to the question will be different for every education jurisdiction, for every duly constituted deliberative group, and very likely for every science teacher (Roberts, 1988, p. 30)

The province of Alberta subsequently regrouped in the early 1990s and adopted a more deliberative process (Roberts, 1995), with the result that their junior and senior high school programs now have an STS science stream that is accepted by the province's universities for students who are not entering science-related fields. Alberta produced its own set of textbooks to support the STS science program in the high school (*Visions I*, *Visions II*, and *Visions III*).

Fensham's (1992) caution about the power of stakeholders was illustrated in Alberta's experience. Fensham warned that some influential stakeholders simply want school science to act as society's screening device to maintain an intellectual, social elite; for example, white male middle-class students have generally enjoyed a privileged status (Lee, 1997; Roth and McGinn, 1998). Reformers must know their own political territory well and must plan ways to negotiate its pathways.

Another group of stakeholders has an interest in maintaining a view of science as: authoritarian, objective, purely rational, nonhumanistic, purely empirical, universal, impersonal, socially sterile, and unencumbered by the vulgarity of human imagination, dogma, judgment, or cultural values (Aikenhead, 1996). Gaskell's (1992) and Gallagher's (1991) research showed that high school science teachers are among the strongest defenders of this view. Thus, it is imperative in the process of deliberation to involve highly credible people (for example, enlightened science teachers and university science professors) who will challenge this stereotype view.

In summary, the first product toward developing an STS curriculum is curriculum policy. STS curriculum policy has a function, content, structure, and sequence, as well as a process for determining that policy. The most promising process is deliberation.

Another product along the road to curriculum development is the material used by classroom teachers.

CLASSROOM MATERIALS AND RESEARCH AND DEVELOPMENT

To meet the demands of STS reform efforts, conscientious teachers require daily professional guidance to help them fulfill the new curriculum policy. These teachers deserve suitable classroom materials (for example, practical teacher guides, booklets, resources, and textbooks). Without suitable materials, an STS science curriculum will not be achieved.

From country to country, cultural conventions differ over how textbooks and materials are developed. The vested interests of the traditional textbook establishment (authors included) can undermine attempts at reform. If STS curriculum developers are to be successful, therefore, alternative processes of developing classroom materials may need to be implemented. The most promising process is R&D.

A short case study will illustrate how R&D can work. This case study, the development of a Canadian grade 10 STS science textbook, will also show how to integrate the processes of deliberation, R&D, and implementation, when producing classroom materials helpful to science teachers.

The textbook, *Logical Reasoning in Science and Technology* (LoRST) (Aikenhead, 1991), evolved from the two separate deliberation processes described previously. These deliberations answered the question, "What counts as science education?" and guided the content, integrative structure, and sequence within LoRST accordingly.

R&D was the process central to producing the textbook. The project was informed by the research literature on student learning, teacher practical knowledge, and STS content itself, and by the developer's earlier experiences producing STS materials (Aikenhead, 1979). The R&D followed a multistage sequence that took place in various classrooms and involved a collaboration with students and teachers (Aikenhead, 1994a).

In the first stage, I wrote and taught the first draft in a local high school. Based on this collaboration, the text was modified to yield the second draft and a rough draft of the teacher guide was written. Students acted as consultants by posing questions out of curiosity, by writing material in response to assignments, by offering advice, and by spontaneously interacting in the classroom. These questions, materials, suggestions, and interactions went into the second draft of LoRST.

In the second stage, this second draft was used by three volunteer teachers who received no inservice preparation but who were capable of being flexible. Their classes were observed daily. This collaboration with teachers and students led to the refinement of the teaching strategies suggested in the teacher guide and led to many revisions in the student materials. Students' language and interactions were incorporated into the text. As a result, LoRST was polished into the third draft, both the student text and teacher guide.

The last stage combined R&D with the *process* of implementation (Table 1)—implementing a new curriculum in the province of Saskatchewan. The implementation process provided a vehicle for obtaining feedback from teachers who were field testing the new curriculum and the third draft of LoRST. In the second stage, the materials had worked well with students. But could the materials work well with a cross section of teachers? The last stage in the R&D process addressed this question. In this implementation process, the 30 teachers became the clients of the R&D project. Teacher feedback helped revise LoRST. The resulting classroom materials were published as a textbook and teacher guide (Aikenhead, 1991), and they are now being used in several provinces across Canada.

Most of the 30 teachers involved in the field testing became responsible for implementing Saskatchewan's science education reform in their own school districts. This implementation process will take many years to complete and will require concerted attention from time to time. I would argue that any implementation is successful if, within five years, 50 percent of the teachers teach science in the way envisaged by the new STSE curriculum policy, and 25 percent will require another five years. For those who will not change, retirement will eventually come.

LoRST, the product of this R&D process, is succinctly described here to illustrate some of the features of STS science mentioned earlier. (For a detailed description of LoRST see Aikenhead, 1992a, 1992b.) LoRST teaches scientific content in conjunction with STS content and critical reasoning skills to a target audience of grade 10 students of average (or above average) academic ability. Students learn scientific facts, concepts, and principles from physics, chemistry, and biology in a way that connects those facts, concepts, and principles with the students' everyday world. The interdisciplinary nature of LoRST places it in category five of the "Categories STS Science" previously described.

The textbook begins with courtroom testimony by scientific experts— a social context familiar to students. This creates the need to know a host of science concepts and logical reasoning skills. In LoRST, the social issue of drinking and driving creates the need to know (1) the technology of the breathalyzer; (2) how science and technology interact with each other, and how they both interact with various aspects of society such as the law; and (3) scientific content such as mixtures, concentration, chemical reactions, photometry, electrical circuits, and the biology of body cells and systems. While the content is "driven by" the social issue of drinking and driving, the content is not limited to that social issue. For instance, students solve concentration problems in the world of recipes, false advertising, toxic chemicals, and farm fertilizers. Classification of mixtures is introduced in the context of the Red Cross and is developed via the technology of salad dressings. Electricity concepts are learned to bridge the gap between atomic theory and the household appliances familiar to adolescents (both female and male). Heat and temperature are taught in an historical context, accompanied by inquiry labs requiring students to construct relevant concepts. The textbook ends with everyday, public policy decisionmaking issues (for example, whether or not to develop an antidrunken driver device for cars). The issue requires students to synthesize the book's scientific and STS content with critical reasoning skills and predispositions. The skill at making different types of decisions (scientific, legal, moral, logical, and public policy decisions) gradually develops with study and practice throughout the book.

LoRST's emphasis on logical reasoning reflects a Canadian curriculum policy to improve students' critical thinking skills (Aikenhead, 1990). Specific critical reasoning skills are taught in Unit 3, "Science & Critical Thinking: The Logic Game." These skills are then applied throughout the book. More important than the individual reasoning skills themselves is the increase in students' *predisposition* (habits of the mind) to analyze, to question, and to articulate a reasoned argument (McPeck, 1981).

This case study of the LoRST project illustrates how the R&D process, in conjunction with the processes of deliberation and implementation, can yield classroom materials that are (1) rationally based in curriculum policy and educational research, and (2) effectively grounded in classroom practice. The R&D study not only focused on the lived experiences of students (giving high priority to the "learned curriculum"), but also collaborated with those students to produce classroom materials in harmony with the intended curriculum, usable by teachers with limited in-service training, and consistent with students' views on relevancy and practical appropriateness. Students contributed significantly to the textbook's content, structure, and language. By engaging students in tasks in the natural setting of their own classroom, I was able to attend to information that spontaneously emerged during instruction or to information that thoughtfully evolved from informal discussions with students.

Another influential STS science project in Canada is *SciencePlus*, developed by the Atlantic Science Curriculum Project (1986) in the maritime provinces and targeted for grades 7–9 (McFadden, 1980). The R&D process that produced these classroom materials (three textbooks) involved teams of classroom teachers coordinated by a university science educator, Charles McFadden (1991). The *SciencePlus* textbook series has been adopted in other Canadian provinces (McFadden *et al.*, 1989), and in the United States (McFadden and Yager, 1997), often after a modification has occurred to match a local curriculum policy. For example, Alberta's STS curriculum led to the production of *SciencePlus Technology and Society* (McFadden *et al.*, 1989), a different textbook for each of grades 7–9. The R&D process that produced *SciencePlus* combined naturally with the process of implementation as will be described.

Up to this point in the chapter, we have examined two pairs of product/process for STS curriculum development—curriculum policy by deliberation and classroom materials by R&D. Two other product/process pairs remain to be addressed, though in much less detail due to space limitations.

TEACHER UNDERSTANDING AND IMPLEMENTATION

Teacher understanding is a major component in the successful development of an STS curriculum. The intended curriculum must be interpreted into the translated curriculum *before* student learning occurs. Teacher understanding is arguably the most influential force in this transformation.

This influence comes to a head in the process of implementation, though it is also prevalent in deliberative inquiries that lead to curriculum policy. In Canada we have not had a systematic study into the implementation of STS curricula across Canada similar to the study conducted by Kumar and Berlin (1996) in the United States. However, several provinces have used various implementation strategies to augment teacher understanding of STS content and its integration with science content (Blades, 1997; Gaskell, 1982; Hart, 1989; Leblanc, 1989; Pedretti and Hodson, 1995; Roberts, 1988).

Science teachers have their own ideas about what constitutes appropriate content, instruction, and assessment. Some teachers' preconceptions will already exemplify the new curriculum policy, but many may not. Teachers' previously held conceptions were constructed during their pre-service education experiences and from their teaching experiences (Aikenhead, 1984; Duffee and Aikenhead, 1992). Their conceptions fulfill many practical purposes, such as coping with, and surviving in, a wide range of classroom contexts and community situations. An inservice program associated with a new STS curriculum is only a tiny increment in a wealth of past experiences that have shaped a teacher's understanding of science teaching. Thus, inservice intervention alone cannot alter a teacher's acceptance of STS science. Teachers' conceptions will not likely change unless those teachers are able to influence their teaching contexts and are able to envision the practical consequences of a new curriculum. This is the common sense reason behind Roberts' (1988) claim that science teachers must be involved in establishing curriculum policy.

Teacher understanding has been the object of a educational research program called "teacher practical knowledge" (Clandinin, 1985; Clandinin and Connelly, 1996; Duffee and Aikenhead, 1992; Lantz and Kass, 1987). Teacher practical knowledge is comprised of many interacting sets of personal ideas, experiences, and feelings of a teacher, including the self-image that a teacher wishes to project. Teacher practical knowledge is *not* pedagogical theory. Its relationship to pedagogical theory is very similar to the relationship between engineering expertise and scientific theory.

By taking on a teacher-practical-knowledge perspective, a curriculum developer pays attention to the common sense inherent in teachers' preconceptions about science teaching, and addresses those common sense preconceptions. (This is similar to a constructivist teacher confronting students' common sense preconceptions about natural phenomena.) For instance, the difficulty in changing from traditional practice to an STS approach was revealed in research by Aikenhead (1984), Gallagher (1985), Lantz and Kass (1987), Mitchner and Anderson (1987), and Olson (1982). Their research offers an alternative view to assuming that teachers are lazy or intransigent, a view often expressed by frustrated curriculum developers when an implementation has failed to take hold. Perhaps the curriculum developers themselves have failed.

The 1985 AETS Yearbook (James, 1985) describes a number of preservice and inservice models for preparing science teachers for STS instruction. Several chapters are devoted to the problems and successes of implementing an STS curriculum in the United States. The AETS volume is an excellent resource for the STS curriculum developer concerned with teacher understanding. The diverse chapters reflect multifaceted problems of teacher understanding and curriculum implementation. Teacher practical knowledge may be a helpful construct to integrate these diverse sets of problems.

I would like to identify one major problem and then suggest some general plans of action that we have found successful in Canadian reform toward STS science. When studying science at university, teachers experience a process of socialization into a discipline (Barnes, 1985; Kuhn, 1970; Ziman, 1984). They then develop deep-seated values about science teaching (Aikenhead, 1984; Pedretti and Hodson, 1995). Many science teachers have been socialized into believing that they too have the responsibility to socialize their students into a discipline (that is, science for the elite, not science for all). Many teachers have the self-image of the "little professor" initiating students into the culture of their scientific discipline. From a teacher's point of view, the best way to initiate students into a discipline is the same way that teacher was initiated (Aikenhead, 1984). STS science, with its goal "science for all," challenges the conventional goal "science for the elite" and the initiation of students into a scientific discipline (Hurd, 1975). STS science often gives lower priority to the curriculum emphasis "solid foundations" than a conventional curriculum does. Therefore, to implement an STS science course successfully is to change the deep-seated, personally cherished values of a number of teachers. Teachers' professional knowledge must go through a Kuhnian paradigm shift. Paradigm shifts are difficult. They involve knowledge, values, assumptions, loyalties, and self-images, and therefore require more than rational arguments and simple inservice programs.

Because science teachers have been socialized by university science professors, then one successful plan of action for achieving reform has been to involve the scientific community—the community responsible for shaping a science teacher's values in the first place, and a community with academic credibility. A cadre of enlightened scientists, carefully selected from industry, government labs, and universities, must relieve science teachers of the burden of socializing students into a scientific discipline. Enlightened scientists (often parents of high school students disenchanted with their science courses) will likely support an STS curriculum policy even more if they were involved in the initial deliberation process for that policy.

Teachers must also add new methods to their repertoire of instructional strategies. A new routine of instruction is best learned from fellow teachers—the people who have practical credibility. A successful plan of action will involve a few cleverly selected teachers chosen to go through an intense inservice experience. They then become inservice leaders in their own regions of the country, passing on their leadership expertise to other teachers who repeat the inservice process in their own communities.

This approach was illustrated with finesse by Leblanc (1989) in a three-year STS inservice project he designed and carried out in the province of Nova Scotia, *prior to* implementing an STS curriculum. He selected teachers who were held in high esteem by their colleagues. A small minority of those teachers were known for having an anti-STS outlook, but they were selected anyway, but on the basis of Leblanc's intuitive expectation that they were open-minded enough to listen to the other teachers and university science professors at the intensive inservice summer programs. Leblanc's patience and planning paid off when Nova Scotia formally implemented an STS science curriculum. He invested three years of inservice work with a small cadre of selected teachers.

Each province in Canada implements STS curriculum in its own way. But the successful cases always targeted *teacher understanding* as the highest priority. Obviously, teacher understanding is enhanced when teachers participate in STS curriculum policy deliberations.

The success of inservice programs is characterized by materials and know-how being passed on from experts to others who work in different locations. Industry calls this method of implementation "technology transfer." Educators could benefit from adopting technology transfer methods from industry. For instance, transfer of expertise requires practical on-site experience and a network of participants. In education this would mean that science teachers who are novices with respect to STS science would spend time in the classroom of an "expert" teacher—one who is implementing an STS course.

Action research is an alternative method. Pedretti and Hodson (1995) conducted a one-year study with six science teachers who were positively predisposed to STS science. The aim was to produce usable curriculum materials through teacher ownership and understanding, all organized around an action research group. Pedretti and Hodson documented teachers' increased understanding in matters of: the nature of science, developing curriculum materials, personal and professional development, and collaboration. In addition, participants reaffirmed many of their personal theories and practices (components of teacher practical knowledge), developed new ones, and had some seriously challenged. This effect was called "reinforcement." The researchers concluded:

> Closely tied to the issue of reinforcement are the increased confidence that the teachers now feel in their personal theories and in their ability to make their own curriculum decisions, and the feelings of enhanced credibility concerning their own educational practice. As a direct consequence of their involvement in the group they now know that they are capable of carrying out meaningful research and contributing to curriculum design and development. (Pedretti and Hodson, 1995, p. 481, emphasis in the original)

The action research method again demonstrated that a combination of "grassroots" and "top-down" approaches to implementation can nurture increased understanding by teachers, an understanding that has personal meaning to their unique teaching situations, and more important, an understanding that has direct implications to their students' learning. A case study of teacher understanding and STS implementation in a grade 3–4 classroom is documented by Pedretti (1997). McFadden's (1991) R&D group that produced Canada's *SciencePlus* series represents a variation to Pedretti and Hodson's method of action research.

The research literature on curriculum implementation is rich in other ideas and schemes to help us plan the process thoughtfully (for example, see Cheek, 1992).

STUDENT LEARNING AND INSTRUCTION–ASSESSMENT

Curriculum policy, classroom materials, and teacher understanding all lead to student learning. Instruction and assessment are the obvious processes that nurture student learning in the formal setting of schools.

Science education research of the 1960s reached an unambiguous conclusion: The classroom teacher will influence student outcomes far more than specific curricula, textbooks, or teaching strategies (Welch, 1969). Thus, student learning from the same STS course can vary significantly from one teacher to another. Within any population of teachers there will be three groups: (1) those whose philosophy of science education is consistent with an STS approach (for example, Pedretti and Hodson's group of teachers); (2) those who are diametrically opposed to an STS approach; and (3) those in the middle who can move in either direction due to persuasion or by the requirement to use certain materials. All three groups will have their own influence on student learning.

Teachers and parents often express the fear that students will not learn as much science content from an STS science curriculum. Their fears are largely unwarranted. Research into student learning shows that spending time on new topics and activities (not normally considered science content but related to that content, for example, STS content) is *not* detrimental to student achievement on traditional science content tests or to careers in science and engineering (Aikenhead, 1994b; Champagne & Klopfer, 1982; Yager & Krajcik, 1989). Therefore, in terms of Roberts' curriculum emphasis "solid foundations" (preparing for the next level of education) described earlier, a high school STS science curriculum will not necessarily be detrimental to student achievement in first year university courses, provided that students have a facility in quantitative problem solving (Aikenhead, 1994b).

STS science instruction has relevance to students' everyday world. Thus, STS instruction tries to make a *real* difference to students' everyday life and to the well being of their community (Solomon and Aikenhead, 1994). While such relevance usually enhances student motivation, and therefore achievement (Mesaros, 1988), relevant contexts may to some extent obfuscate the acquisition of science content and the solving of science problems (Solomon, 1987). Students tend to experience difficulty when moving between the theoretical world of pure science concepts, characterized by logical reasoning with evidence, and their everyday world of common sense concepts, characterized by social interactions and consensus (Hennessy, 1993; Lijnse, 1990). If STS science requires students to learn the science content in enough depth to use in everyday situations (rather than to memorize for an examination), then STS science has taken on a *much more rigorous task* than traditional science. This in-depth learning contrasts with making a *political* difference to students' lives by passing tests that artificially open doors to social opportunities (for example, attending a university), but without achieving any meaningful learning of the science content (Costa, 1997).

Because STS instruction aims to make a real difference to a student's everyday life, STS science educators run the risk of judging their own success by much higher standards and expectations than teachers who subscribe to the standard of getting students through their course or catering to the elite students who have the savvy to learn meaningfully on their own. In this sense, then, traditional science instruction-assessment can be viewed as "soft" and superficial while STS science instruction-assessment can be thought of as "hard" and rigorous. For instance, memorizing how to solve heat transfer problems is superficial. Explaining how the conceptual invention of energy changed scientists' ideas about heat transfer, on the other hand, is rigorous. The assessment of student learning can be superficial or rigorous.

The problem of superficial learning was dramatically discovered by Larson (1995) when she found students in a high school chemistry class who actually told her the rules they followed so they could pass Mr. London's chemistry class without really understanding much of chemistry. Larson called these rules "Fatima's rules" after the most articulate student in the class. For example, one rule was not to read the textbook but to memorize the bold face words and phrases. Fatima's rules include such coping or passive-resistance mechanisms as "silence, accommodation, ingratiation, evasiveness, and manipulation" (Atwater, 1996, p. 823). What results is not meaningful learning but merely "communicative competence" (Kelly and Green, 1998) or "an accoutrement to specific rituals and practices of the science classroom" (Medvitz, 1996, p. 5). Loughran and Derry (1997) investigated students' reactions to a science teacher's concerted effort to teach for meaningful learning ("deep understanding") as STS science teachers do. The researchers found a reason for Fatima's rules, a reason related to the culture of public schools:

> The need to develop a deep understanding of the subject may not have been viewed by them [the students] as being particularly important as progression through the schooling system could be achieved without it. In this case such a view appears to have been very well reinforced by Year 9. This is not to suggest that these students were poor learners, but rather that they had learnt how to learn sufficiently well to succeed in school without expending excessive time or effort. (Loughran and Derry, 1997, p. 935)

Their teacher lamented, "No matter how well I think I teach a topic, the students only seem to learn what they need to pass the test, then, after the test, they forget it all anyway" (Loughran and Derry, 1997, p. 925). Tobin and McRobbie (1997, p. 366) documented a teacher's complicity in Fatima's rules: "There was a close fit between the goals of Mr. Jacobs and those of the students and satisfaction with the emphasis on memorisation of facts and procedures to obtain the correct answers needed for success on tests and examinations." When playing Fatima's rules, students (and some teachers) make it appear as though meaningful learning has occurred, but at best rote memorization of key terms and processes is only achieved temporarily.

Costa (1997) synthesized the work of Larson (1995) and Tobin and McRobbie (1995) with her own classroom research and concluded:

> Mr. Ellis' students, like those of Mr. London and Mr. Jacobs, are not working on chemistry; they are working to get through chemistry. The subject does not matter. As a result, students negotiate treaties regarding the kind of work they will do in class. Their work is not so much productive as it is political. They do not need to be productive—as in learning chemistry. They only need to be political—as in being credited for working in chemistry. (Costa, 1997, p. 1020)

The three teachers (Ellis, London, and Jacobs) exemplify the superficial teaching that can pass as legitimate instruction in traditional classes. But superficial teaching can become obviously transparent in an STS science class.

The main point is this: the general goal "science for all" associated with STS learning in Canada represents a *political* paradigm shift from the traditional goal "science for the elite." Learning and instruction–assessment will change accordingly, if STS curriculum development is to succeed.

Today we recognize that learning will likely be more effective when classroom activities serve both instruction and assessment functions (Black, 1997; Gallagher *et al.*, 1996). As a result, formative assessment techniques that accumulate data while instruction takes place (for example, quizzes, check lists, portfolios, concept maps, posters, and self-assessments) are conceived to be *instructional* strategies as well as assessment techniques. In the classroom, instruction and assessment are best integrated. However, when discussing the two processes, it will be convenient to separate the two.

Instruction

Traditional science teaching methods tend to be characterized by convergent thinking and lecture–demonstrations. STS science instruction, however, includes divergent thinking but demands a wider repertoire of teaching strategies (Solomon and Aikenhead, 1994).

Instructional strategies for STS science were first addressed systematically in 1980 by Ziman in his book *Teaching and Learning about Science and Society*. Solomon's (1993) *Teaching Science, Technology and Society* is an excellent current resource for technology transfer programs for STS science teachers. Aikenhead (1988) developed a monograph and videotapes, as part of Saskatchewan science reform, to show how to use specific STS instructional strategies. This monograph, *Teaching Science Through a Science-Technology-Society-Environment Approach: An Instructional Guide*, gives special attention to instructional methods that produce interactivity among students, for instance, divergent thinking, small group work, student-centered class discussion, problem-solving, simulations, decisionmaking, controversies, debating, and using the media and other community resources. In addition, the teacher guide that accompanies the STS textbook *Logical Reasoning in Science & Technology* (Aikenhead, 1991) coaches teachers through activities that work best if they use student interactivity.

Aikenhead (1994b) reviewed the research literature on STS instruction and found little research identifying the effects of STS teaching methods. Notable exceptions included the Discussion of Issues in School Science (DISS) project, a research program based on small-group work and applied specifically to STS science content (Solomon, 1988). The DISS project documented students' capabilities at conducting effective small-group discussions on science-related social issues. Byrne and Johnstone (1988) generalized the efficacy of small-group discussions. They concluded: "It is the achievement of interactivity, rather than the exact format, whether it be simulation, group discussion or role playing" (p. 44). Interactive learning approaches are often identified with STS science instruction. The research evidence suggests the following (Byrne and Johnstone, 1988):

1. Simulations and educational games can be just as effective as traditional methods in teaching science content. Simulations and games can be far more effective than traditional methods in helping to develope positive attitudes
2. In terms of attitude development, the strategies of role playing, discussion, and decisionmaking can be *highly* effective
3. "Group discussion can stimulate thought and interest and develop greater commitment on the part of the student." (p. 45)
4. An analysis and evaluation of historical case studies can be effective in promoting an understanding of the processes of science.

These findings were supported by the R&D project that produced *Logical Reasoning in Science* and *Technology* (Aikenhead, 1991) described earlier. According to 80 percent of the students who helped develop the third draft of the textbook, simulations served as concrete connections between the everyday world and academic science content, and simulations made the academic science more interesting to learn. Only 8 percent of the students found simulations of little or no value.

In general, taking on STS instructional methods usually involves a professional paradigm shift in teachers' ideas *away from* a scientist-dominated view of the world conveyed to students by a teacher-centered approach to teaching, *toward* a student-dominated view of the world (informed by science and technology) conveyed by more student-centered approaches to teaching.

Assessment

The process of instruction and the product of student learning are intricately tied to the process of assessment. Therefore, good assessment is indistinguishable from good instruction.

The professional and political paradigm shifts associated with STS instruction have direct implications for assessment practices beyond the assessment of students. In Canada, we are trying to broaden assessment to include two major issues: assessing the new STS curricula themselves, and assessing the support experienced by teachers. Part of the design of an STS curriculum should include concrete plans for assessing the support that teachers receive from government, industry, universities, and parents. The stakeholders who participated in the deliberation process must initiate ways by which the jurisdictions they represent will be evaluated by teachers in terms of support during an extended implementation process. Stakeholders must be held accountable to teachers in any reform effort, and this accountability must be designed into the STS curriculum policy from the very beginning.

For the purpose of clear and critical thinking, it is important to distinguish between the act of observing or collecting student work and the act of interpreting or judging that collection of work. However, confusion arises over what to name each act. Some educators, by convention, use the same term for both acts. This would tend to equate the methods of collecting data with the methods of interpreting those data. (In science classes, we usually teach students the difference between observing and interpreting.) Recently in Canada, we have begun to refer to the act of collecting student work as "*assessment*" and the act of judging that work as "*evaluation*." This distinction can help teachers escape the old paradigm of student assessment associated with traditional science. For instance, teachers can practice new *assessment* techniques without necessary changing their *evaluation* standards.

The challenge of such a paradigm shift was clarified in Canada by Ryan (1988) when he described three paradigms of assessment and evaluation (based on the work of Habermas, 1971):

1. *Empirical–analytic*—Western technical rationalism embodied in logical positivist origins. This amounts to the traditional standardized approach to assessment and evaluation.
2. *Interpretive*—Understanding student's language, concepts, and actions from the student's point of view. Alternative assessment techniques, such as portfolios and concept mapping, illustrate this paradigm.
3. *Critical–theoretic*—The elimination of oppressive human relationships (oppressive is defined as forced assimilation). Two examples would be: assessment rubrics for thoughtful decisionmaking developed collaboratively between teacher and students, and student self-evaluation.

These three paradigms clarify key issues in assessment and evaluation: (1) the issue of standardized tests falls within the empirical–analytic paradigm;

(2) the issue of formative assessment is clearly within the interpretive paradigm; and (3) the issues of equity and student empowerment both fit primarily within the critical–theoretic paradigm. We must be able to function eclectically in all three paradigms. For the sake of clear thinking, however, we should not lose sight of the paradigm we are in, at any given moment.

For example, there are many purposes of assessment and evaluation in Canadian schools. Student certification and the development of educational policy are treated within the empirical–analytic paradigm. Teaching improvement usually falls within the interpretive paradigm. Student empowerment is clearly within the critical–theoretic paradigm.

Student outcomes also vary, so they require different approaches to assessment and evaluation depending on the appropriate paradigm. The empirical–analytic paradigm focuses on the *product* of instruction and the students's tangible work, and it gives priority to the quantitative standardization of that work. The interpretive paradigm focuses on both the student's product and on how the student produced the work—the *process*. It embraces nonquantitative assessment techniques, such as rubrics, concept mapping, check lists, and authentic assessment (Black, 1993, 1997). The critical–theoretic paradigm gives special attention to the social or cultural *context* in which assessment takes place, a context that greatly influences both the process and product of a student's work (Roth and McGinn, 1998). The critical–theoretic paradigm focuses on the product, process, and context of student learning.

The assessment and evaluation of scientific literacy has traditionally been conducted well within the empirical-analytic paradigm (Aikenhead, Fleming and Ryan, 1987). Student responses are either right or wrong, with little interpretation and little consideration for context. Scores are standardized against such norms as statistical distributions or judgments by panels of experts. Assessment tends to be confounded with evaluation, thereby merging the two concepts into one. An alternative type of instrument that operates within the interpretive paradigm for assessing STS content is *Views on Science-Technology-Society*, VOSTS (Aikenhead, Ryan, and Fleming, 1989). By collaborating with students, the researchers developed an empirically based, multiple-choice, assessment instrument (Aikenhead and Ryan, 1992). The empirical data consisted of the students' written and oral work itself. As a consequence, students are generally able to express their personal and reasoned viewpoints in their own language when they respond to any of the 114 VOSTS items. VOSTS can serve as a point of departure for formulating an item bank unique to any STS science curriculum.

As Champagne and Newell (1992) point out, certain educational jurisdictions demand that assessment be simplistic, competitive, and unidimensional in order to distinguish winners from losers. Tests are designed "on the assumption that knowledge can be represented by an accumulation of bits of information [playing Fatima's rules] and that there is one right answer" (p. 846). On the other hand, "alternative assessment is based on the assumption that knowledge is actively constructed by the child and varies from one context to another" (p. 847). We can now identify these two positions as exemplifying the empirical-analytic and interpretive paradigms, respectively. Moreover, using the three-paradigm framework suggested by Ryan (1988), we can now ask ourselves the question, What does the critical-theoretic paradigm say about what knowledge is important to learn? The answer leads to other issues such as: Whose knowledge is privileged in the assessment? Whose social interactions have cultural capital? Whose goals define the criteria for evaluation and how are these goals established? These critical-theoretic issues are discussed by O'Loughlin (1992).

My R&D project that produced *Logical Reasoning in Science & Technology*, LoRST (described previously) uncovered the need to guide teachers as they attempted teaching new STS science content using alternative methods of instruction and using new assessment and evaluation techniques. Throughout the teacher guide to LoRST, there are teacher inservice sections on assessment and evaluation. This includes sample work from a cross-section of students, along with how teachers assessed the work, and how the work can be evaluated. Concept maps, check lists, and rubrics are featured, although postermaking and portfolios are also mentioned.

Conclusion

The STS science curricula developed in Canadian provinces all deal with instruction and assessment–evaluation in their own particular way. Some provinces (such as Saskatchewan) have produced their own publications which were originally used with inservice programs, but are now found in preservice B.Ed. programs at universities. The processes of instruction and assessment–evaluation were beyond the scope of the *Common Framework of Science Learning Outcomes* (CMEC, 1997). It devotes most of its attention to specifying the student learning (organized within four areas: STSE content, skills, canonical science knowledge, and attitudes) expected in a Canadian STSE science curriculum. The processes of achieving that learning is left to the individual provinces.

SUMMARY AND IMPLICATIONS

This chapter explored four pairs of fundamental processes and crucial products that lead to a successful STS science curriculum. Each of the crucial products—curriculum policy, classroom materials, teacher understanding, and student learning—was associated with a different process—deliberation, R&D, implementation, and instruction–assessment, respectively. These products and processes are interrelated as depicted in Table 1 by the designation of "high" and "low" associations between them.

This chapter covered the territory that educators should explore if they expect to develop successful STS science curricula. Important features of this territory were illustrated by Canadian examples. Although regions within the United States (or in any other country) will have their own unique features, the coordination of each of the four product–process pairs identified in this chapter will be essential to producing successful STS teaching in any country.

Based on our Canadian experiences with STS science teaching, science educators in other countries can better anticipate the nature of the challenges they face as they work with students, teachers, and other stakeholders, many of whom will need to experience a type of paradigm shift. The biggest challenge in Canada continues to be the paradigm shift over who has privilege and the political power to decide what ought to be learned in science classes and what ideologies will prevail—"science for all" or "science for an elite"? (Blades, 1997; Fensham, 1992; Roberts, 1988). Some regions in the United States will have the issue of privilege and political power well established by cultural convention, while in other regions the issue will need to be negotiated. From region to region the political topography will vary according to the dominant stakeholders, but the tension between "science for all" and "science for an elite" can be anticipated, along with other tensions and dilemmas described in this chapter.

The appropriate way to deliberate over curriculum policies will depend on the cultural convention of that region, but deliberations must take place nevertheless. When contemplating how to engage in the process of deliberation, a science educator must know the political topography of the community. This savvy will be essential to success. National projects such as *Project 2061* (AAAS, 1989) and *Standards* (NRC, 1996) can influence the political topography. These two projects do address the ideological choice between "science for all" and "science for an elite," however, both projects appear to emphasize science for an elite because they privilege the assimilation of students into thinking like scientists think (Aikenhead, 1997). The political self-interest of all stakeholders will pervade deliberations and curriculum policy in any country.

In this chapter we discovered political power is an issue in both teacher understanding and student learning. It was recognized as a focus within the critical-theoretic paradigm of assessment/evaluation. How it plays out in a non-Canadian setting will need to be carefully considered by STS science educators wishing to reform their own science curriculum.

The territory mapped out in this chapter embraced the intended, the translated, and the learned curriculum. All three must be considered by any educational reform. To clarify the nature of those curricula, we found it very helpful to consider the concept called "curriculum emphases" and we found it important to identify those emphases in all the products of curriculum change: curriculum policy, teaching materials, teacher understanding, and student learning. Although the *priority* of curriculum emphases chosen for a given region will vary, what is crucial is the need to identify this priority. Unarticulated priorities and curriculum emphases lead to failed reform efforts.

STS *content* constituted another fundamental feature to the territory mapped out in this chapter. Similar to curriculum emphases, STS content will vary from region to region. However, the *structure* for integrating science content with STS content, identified in the chapter by an eight-category scheme, will be an essential feature to any STS science instruction world wide. STS educators are well advised to use the scheme to reflect on their own views and to communicate those views to teachers and other stakeholders.

Teacher understanding was clarified in Canada through "teacher practical knowledge," which includes the teacher's past experiences of being socialized into a scientific discipline. For reform minded science educators, one implication to this entrenched social practice is to initiate counter measures. Technology transfer and collaborative action research both hold promise to supplement the normal implementation methods used in any community.

And last, we explored the tensions between meaningful learning and superficial learning: students' playing Fatima's rules. By playing Fatima's rules, we privilege "science for an elite" because our instruction tends to screen out those students who do not share the worldview embraced by most scientists (Cobern, 1991; O'Loughlin, 1992). Screening out students may be a cultural feature of a community's schooling. When STS science reform is contemplated for such a community, the dilemma over meaningful versus superficial learning will emerge. Those science educators familiar with a constructivist approach to meaningful learning (Tobin, 1993) will have already experienced reactions against this innovation and the enormous pressures placed on teachers to play Fatima's rules (under the guise of "covering the curriculum" sanctified by scientific authorities, including national associations and councils). Costa's (1997) and Loughran and Derry's (1997) recent work spoke to those pressures.

In a recent comparative study of national science curricula in United States, Australia, New Zealand, England and Wales, and the province of Ontario, Orpwood and Barnett (1997) reinforced the need to support the development of teachers as they attempt to implement reform. Orpwood and Barnett concluded:

> In our view, a curriculum framework is a necessary but not a sufficient condition for quality teaching and learning. Teachers' use of a curriculum framework in the classroom to serve the increasingly diverse needs of their pupils requires considerable sophistication of understanding and professional judgement and creativity. No national framework or curriculum policy, however well developed, can function as a direct instrument to affect pupils' learning—the failure of past attempts at "teacher-proof curricula" are surely evidence of this. What a framework can provide, however, is a common set of goals and expectations from which teachers can design programmes which are meaningful and effective in their specific contexts. (p. 347)

This chapter has described myriad other products and processes (beyond curriculum policy and teacher understanding) that need to be orchestrated to create a meaningful and effective STS science experience for students.

The professional and political paradigm shifts associated with successful STS reform have direct implications for assessment and evaluation practices. In addition to the implications for student assessment discussed in the chapter, we need to broaden those implications to include *evaluating the support experienced by teachers*. Part of the design of a rational reform effort should include concrete plans for assessing the support that teachers receive from government, industry, universities, and parents. The stakeholders who participated in the deliberation process over STS reform must initiate ways by which the jurisdictions they represent will be evaluated *by teachers* in terms of the perceived support provided during an extended implementation process. Stakeholders must be held accountable to teachers in any reform effort, and this accountability must be designed into the curriculum policy from the very beginning.

REFERENCES

AAAS. (1989). *Project 2061: Science for all Americans*, American Association for the Advancement of Science, Washington, D.C.

Aikenhead, G. S. (1979). Science: A way of knowing. *The Science Teacher* 46(6):23–25.

Aikenhead, G. S. (1980). *Science in social issues: Implications for teaching.* Science Council of Canada, Ottawa, Canada.

Aikenhead, G. S. (1984). Teacher decision making: The case of Prairie High. *Journal of Research in Science Teaching* 21:167–186.

Aikenhead, G. S. (1985). Collective decision making in the social context of science. *Science Education* 69:453–475.

Aikenhead, G. S. (1988). *Teaching Science through a Science-Technology-Society-Environment Approach: An Instruction Guide*, SIDRU, Faculty of Education, University of Regina, Regina, Saskatchewan.

Aikenhead, G. S. (1990). Scientific/technological literacy, critical thinking, classroom practice. In Norris, S. and Phillips, L. (eds.), *Foundations of Literacy Policy in Canada*, Detselig Enterprises, Alberta, pp. 127–145.

Aikenhead, G. S. (1991). *Logical reasoning in science and technology.* John Wiley of Canada, Toronto.

Aikenhead, G. S. (1992a). Logical reasoning in science and technology. *Bulletin of Science, Technology and Society* 12:149–159.

Aikenhead, G. S. (1992b). The integration of STS into science education. *Theory into Practice* 31(1):27–35.

Aikenhead, G. S. (1994a). Collaborative research and development to produce an STS course for school science. In Solomon, J. and Aikenhead, G. (eds.), *STS education: International Perspectives on Reform*, Teachers College Press, New York. pp. 216–227.

Aikenhead, G. S. (1994b). Consequences to learning science through STS: A research perspective. In Solomon, J. and Aikenhead, G. (eds.), *STS Education: International Perspectives on Reform*, Teachers College Press, New York, pp. 169–186.

Aikenhead, G. S. (1994c). The social contract of science: Implications for teaching science. In Solomon, J. and Aikenhead, G. (eds.), *STS Education: International Perspectives on Reform* Teachers College Press, New York, pp. 11–20.

Aikenhead, G. S. (1994d). What is STS teaching? In Solomon, J. and Aikenhead, G. (eds.), *STS Education: International Perspectives on Reform*, Teachers College Press, New York, pp. 47–59.

Aikenhead, G. S. (1996). Science education: Border crossing into the subculture of science. *Studies in Science Education* 27:1–51.

Aikenhead, G. S. (1997). Toward a First Nations cross-cultural science and technology curriculum. *Science Education* 81:217–238.

Aikenhead, G. S., Fleming, R. W., and Ryan, A. G. (1987). High school graduates' beliefs about science-technology-society. Part I. Methods and issues in monitoring students views. *Science Education* 71:145–161.

Aikenhead, G. S., and Ryan, A. G. (1992). The development of a new instrument: Views on science-technology-society (VOSTS). *Science Education* 76:477–491.

Aikenhead, G. S., Ryan, A., and Fleming, R. (1989). *Views on Science-Technology-Society* (form CDN.mc.5), Department of Curriculum Studies, University of Saskatchewan, Saskatoon, Canada.

American Chemical Society (1988). *ChemCom: Chemistry in the Community*, Kendall/Hunt, Dubuque, Iowa.

Atlantic Science Curriculum Project (1986). *SciencePlus*, Harcourt Brace Jovanovich, Toronto.

Atwater, M. M. (1996). Social constructivism: Infusion into the multicultural science education research agenda. *Journal of Research in Science Teaching* 33:821–837.

Barnes, B. (1985). *About Science*, Basil Blackwell, Oxford.

Bingle, W. H., and Gaskell, P. J. (1994). Scientific literacy for decision making and the social construction of scientific knowledge. *Science Education* 78:185–201.

Black, P. (1993). Formative and summative assessment by teachers. *Studies in Science Education* 21:49–97.

Black, P. (1997). Assessment in the service of science education. In KEDI (Korean Educational Development Institute), *Globalization of Science Education*. Pre-conference proceedings for the International Conference on Science Education, Seoul, Korea, pp. 165–177.

Blades, D. (1997). *Procedures of Power and Curriculum Change*, Peter Lang, New York.

Bybee, R. W. (1985). The Sisyphean question in science education: What should the scientifically and technologically literate person know, value and do—as a citizen? In Bybee, R. W. (ed.), *Science-Technology-Society*, 1985 NSTA Yearbook National Science Teachers Association, Washington, D.C., pp 79–93.

Bybee, R. W., and Mau, T. (1986). Science and technology related to global problems: An international survey of science educators. *Journal of Research in Science Teaching* 23:599–618.

Byrne, M. S., and Johnstone, A. H. (1988). How to make science relevant. *School Science Review* 70(251):43–46.

Champagne, A. B., and Klopfer, L. E. (1982). A causal model of students' achievement in a college physics course. *Journal of Research in Science Teaching* 19:299–309.

Champagne, A. B., and Newell, S. T. (1992). Directions for research and development: Alternative methods of assessing scientific literacy. *Journal of Research in Science Teaching* 29:841–860.

Cheek, D. W. (1992). *Thinking Constructively about Science, Technology and Society Education*, SUNY Press, Albany, NY.

Clandinin, D. J. (1985). Personal practical knowledge: A study of teachers' classroom images. *Curriculum Inquiry* 14:361–385.

Clandinin, D. J., and Connelly, F. M. (1996). Teachers' professional knowledge landscapes: Teacher stories—stories of teachers—school stories—stories of schools. *Educational Research* 25(3):24–30.

CMEC (1997). *Common Framework of Science Learning Outcomes*, Council of Ministers of Education of Canada, Ottawa, Canada.

Cobern, W. W. (1991). *World View Theory and Science Education Research*, NARST Monograph No. 3. National Association for Research in Science Teaching, Manhattan, KS.

Costa, V. (1997). How teacher and students study "all that matters" in high school chemistry. *International Journal of Science Education* 19:1005–1023.

Cross, R. T., and Price, R. F. (1992). *Teaching Science for Social Responsibility*, St. Louis Press, Sydney.

Duffee, L., and Aikenhead, G. S. (1992). Curriculum change, student evaluation, and teacher practical knowledge. *Science Education* 76:493–506.

Eijkelhof, H. M. C. (1994). Toward a research base for teaching ionizing radiation in a risk perspective. In Solomon, J. and Aikenhead, G. (eds.), *STS Education: International Perspectives on Reform*, Teachers College Press, New York, pp. 205–215.

Eijkelhof, H. M. C., and Kortland, K. (1987). Physics in its personal, social and scientific context. *Bulletin of Science, Technology and Society* 7:125–136.

Eijkelhof, H. M. C., and Kortland, K. (1988). Broadening the aims of physics education. In Fensham, P. J. (ed.), *Development and Dilemmas in Science Education*, Falmer Press, New York, pp. 285–305.

Fensham, P. J. (1988). Approaches to the teaching of STS in science education. *International Journal of Science Education* 10:346–356.

Fensham, P. J. (1992). Science and technology. In Jackson, P. W. (ed.), *Handbook of Research on Curriculum*. Macmillan, New York, pp. 789–829.

Fensham, P. J. (1993). Reflections on science for all. In Whitelegg, L. (ed.), *Challenges and Opportunities in Science Education*. The Open University, Milton Keynes, UK.

Gallagher, J. J. (1985). Secondary science teaching practices with implications for STS implementation. In James, R. K. (ed.), *Science, Technology and Society: Resources for Science Educators*, 1985 AETS Yearbook. SMEAC Center, Ohio State University, Columbus, Ohio, pp. 23–32.

Gallagher, J. J. (1991). Prospective and practicing secondary school science teachers' knowledge and beliefs about the philosophy of science. *Science Education* 75:121–133.

Gallagher, J. J., Parker, J., Alvarado, T., Haley, G., and Sandys, D. (1996). Initiating and maintaining continuous assessment in middle school science classes. A paper presented at the annual meeting of the National Association for Research in Science Teaching, St. Louis, MO, March 31–April 3, 1996.

Gaskell, J. P. (1982). Science, technology and society: Issues for science teachers. *Studies in Science Education* 9:33–46.

Gaskell, J. P. (1989). Science and technology in British Columbia: A course in search of a community. *Pacific Education* 1(3):1–10.

Gaskell, P. J. (1992). Authentic science and school science. *International Journal of Science Education* 14:265–272.

Gilliom, M. E., Helgeson, S. L., and Zuga, K. F. (eds.) (1991). *Theory into Practice* (special issue on STS), 30(4):230–315.

Gilliom, M. E., Helgeson, S. L., and Zuga, K. F. (eds.) (1992). *Theory into Practice* (special issue on STS), 31(1):2–84.

Hansen, K.-H., and Olson, J. (1996). How teachers construe curriculum integration: The science, technology, society (STS) movement as Bildung. *Journal of Curriculum Studies* 28:669–682.

Hart, E. P. (1989). Toward renewal of science education: A case study of curriculum policy development. *Science Education* 73:607–634.

Habermas, J. (1971). *Knowledge and Human Interest*, Beacon, Boston.

Hennessy, S. (1993). Situated cognition and cognitive apprenticeship: Implications for classroom learning. *Studies in Science Education* 22:1–41.

Holton, G., Rutherford, J., and Watson, F. (1970). *The Project Physics Course*, Holt, Rinehart and Winston, New York.

Hunt, J. A. (1988). SATIS approaches to STS. *International Journal of Science Education* 10:409–420.

Hurd, P. (1975). Science, technology and society: New goals for interdisciplinary science teaching. *The Science Teacher* 42:27–30.

Hurd, P. (1989). Science education and the nation's economy. In Champagne, A. B., Lovitts, B. E., and Calinger, B. J. (eds.), *Scientific Literacy*, American Association for the Advancement of Science, Washington, D.C., pp. 15–40.

James, R. K. (ed.) (1985). *Science, Technology and Society: Resources for Science Educators*, AETS Yearbook. SMEAC Center, Ohio State University, Columbus, Ohio.

Keeves, J., and Aikenhead, G. S. (1995). Science curricula in a changing world. In Fraser, B. J. and Walberg, H. J. (eds.) *Improving Science Education*, The National Society for the Study of Education, Chicago, pp. 13–45.

Kelly, G. J., and Green, J. (1998). The social nature of knowing: Toward a sociocultural perspective on conceptual change and knowledge construction. In Guzzetti, B. and Hynd, C. (eds.), *Perspectives on Conceptual Change*, Lawrence Erlbaum, Mahwah, NJ.

Kranzberg, M. (1991). Science-technology-society: It's as simple as XYZ! *Theory in Practice* 30:234–241.

Kuhn, T. (1970). *The Structure of Scientific Revolutions* (2nd ed.), University of Chicago Press, Chicago.

Kumar, D., and Berlin, D. (1996). A study of STS curriculum implementation in the United States. *Science Educator* 5(1):12–19.

Lantz, O., and Kass, H. (1987). Chemistry teachers' functional paradigms. *Science Education* 71:117–134.

Larson, J. O. (1995). Fatima's rules and other elements of an unintended chemistry curriculum. Paper presented at the American Educational Research Association annual meeting, San Francisco, April 18–22, 1995.

Layton, D. (1994). STS in the school curriculum: A movement overtaken by history? In Solomon, J. and Aikenhead, G. (eds.), *STS Education: International Perspectives on Reform*, Teachers College Press, New York, pp. 32–44.

Leblanc, R. (1989). *Department of education summer science institute*. Halifax, Canada: PO Box 578.

Lee, O. (1997). Scientific literacy for all: What is it, and how can we achieve it? *Journal of Research in Science Teaching* 34:219–222.

Lijnse, P. (1990). Energy between the life-world of pupils and the world of physics. *Science Education* 74:571–583.

Loughran, J. and Derry, N. (1997). Researching teaching for understanding: The students' perspective. *International Journal of Science Education* 19:925–938.

McFadden, C. (1980). Barriers to science education improvement in Canada: A case in point. In McFadden, C. (ed.), *World Trends in Science Education* Atlantic Institute of Education, Halifax, Canada, pp. 49–59.

McFadden, C. (1991). Toward an STS school curriculum. *Science Education* 75:457–469.

McFadden, C., and Yager, R. (1997). *SciencePlus Technology and Society*, Holt, Rinehart & Winston, Austin, TX.

McFadden, C., Morrison, E. S., Armour, N., Moore, A., Hammond, A. R., Nicoll, E. M., Haysom, J., and Smyth, M. M. (1989). *SciencePlus Technology and Society*, Harcourt Brace Jovanovich, Toronto.

McGinn, R. E. (1991). *Science, Technology, and Society*, Prentice Hall, Englecliffs, NJ.

McPeck, J. E. (1981). *Critical Thinking and Education*, Martin Robertson Press, Oxford.

Medvitz, A. G. (1996). *Science, schools and culture: The complexity of reform in science education*. A paper presented to the 8th symposium of the International Organization for Science and Technology Education (IOSTE), Edmonton, Canada, August 17–22, 1996.

Mesaros, R. A. (1988). The effect of teaching strategies on the acquisition and retention of knowledge. (Doctoral dissertation, University of Pennsylvania, 1987). *Dissertation Abstracts International* 49(2):229–A.

Mitchener, C., and Anderson, R. (1989). Teachers' perspective: Developing and implementing an STS curriculum. *Journal of Research in Science Teaching* 24:351–369.

Nagasu, N., and Kumano, Y. (1997). *STS initiatives in Japan*. A paper presented to the International Conference on Science Education: Globalization of Science Education, Korean Education Development Institute, Seoul, Korea, May 26–30, 1997.

National Research Council (NRC). (1996). *National Science Education Standards*, National Academy Press, Washington, D.C.

O'Loughlin, M. (1992). Rethinking science education: Beyond Piagetian constructivism toward a sociocultural model of teaching and learning. *Journal of Research in Science Teaching* 29:791–820.

Olson, J. (ed.) (1982). *Innovation in the Science Curriculum*, Croom Helm, London.

Ogawa, M. (1995). Science education in a multi-science perspective. *Science Education* 79:583–593.

Orpwood, G. (1985). Toward the renewal of Canadian science education. I. Deliberative inquiry model. *Science Education* 69:477–489.

Orpwood, G., and Barnett, J. (1997). Science in the national curriculum: An international perspective. *The Curriculum Journal* 8:331–349.

Oxford University Department of Educational Studies (1989). *Enquiry into the Attitudes of Sixth-formers towards Choice of Science and Technology Courses in Higher Education*, Department of Educational Studies, Oxford.

Pedretti, E. (1997). Septic tank crisis: A case study of science, technology and society education in an elementary school. *International Journal of Science Education* 19:1211–1230.

Pedretti, E., and Hodson, D. (1995). From rhetoric to action: Implementing STS education through action research. *Journal of Research in Science Teaching* 32:463–485.

Piel, E. P. (1981). Interaction of science, technology, and society in secondary schools. In Harms, N. and Yager, R. (eds.), *What Research Says to the Science Teacher*, Volume 3, National Science Teachers Association, Washington, D.C., pp. 94–112.

Roberts, D. A. (1982). Developing the concept of "curriculum emphases" in science education. *Science Education* 66:243–260.

Roberts, D. A. (1983). *Scientific Literacy*, Science Council of Canada, Ottawa, Canada.

Roberts, D. A. (1988). What counts as science education? In Fensham, P. J. (ed.), *Development and Dilemmas in Science Education*. Falmer Press, New York, pp. 27–54.

Roberts, D. A. (1995). Junior high school science transformed: Analysing a science curriculum policy change. *International Journal of Science Education* 17:493–504.

Rosenthal, D. B. (1989). Two approaches to STS education. *Science Education* 73:581–589.

Roth, W-M., and McGinn, M. K. (1998). unDELETE lives, work, and voices. *Journal of Research in Science Teaching*, 35:399–421.

Ryan, A. G. (1988). Program evaluation within the paradigm: Mapping the territory. *Knowledge: Creation, Diffusion, Utilization* 10(1):25–47.

SCC (1984). *Science for Every Student: Educating Canadians for Tomorrow's World* (Report No. 36). Science Council of Canada, Ottawa, Canada.

Schwab, J. J. (1974). Decision and choice: The coming duty of science teaching. *Journal of Research in Science Teaching* 11:309–317.

Schwab, J. J. (1978). *Science, Curriculum, and Liberal Education*, University of Chicago Press, Chicago.

Sjøberg, S. (1996). *Scientific literacy and school science: Arguments and second thoughts*. A paper presented at the seminar on Science, Technology and Citizenship, Leangkollen, Oslo, Norway, November 22–24, 1996.

Snow, R. E. (1987). Core concepts for science and technology literacy. *Bulletin of Science, Technology and Society* 7:720–729.

Solomon, J. (1981). Science and society studies in the curriculum. *School Science Review* (82):213–220.

Solomon, J. (1987). Social influences on the construction of pupil's understanding of science. *Studies in Science Education* 14:63–82.

Solomon, J. (1987). Social influences on the construction of pupil's understanding of science. *Studies in Science Education* 14:63–82.

Solomon, J. (1988). Science technology and society courses: Tools for thinking about social issues. *International Journal of Science Education* 10:379–387.

Solomon, J. (1993). *Teaching Science, Technology and Society*. Open University Press, Buckingham, UK.

Solomon, J. (1994). Conflict between mainstream science and STS in science education. In Solomon, J. and Aikenhead, G. (eds.), *STS Education: International Perspectives on Reform* Teachers College Press, New York, pp. 3–10.

Solomon, J., and Aikenhead, G. (eds.) (1994). *STS Education: International Perspectives on Reform*. Teachers College Press, New York.

Tobin, K. (ed.) (1993). *The Practice of Constructivism in Science Education*, American Association for the Advancement of Science, Washington, D.C.

Tobin, K., and McRobbie, C. (1996). Cultural myths as constraints to the enacted science curriculum. *Science Education* 80:223–241.

Tobin, K., and McRobbie, C. (1997). Beliefs about the nature of science and the enacted science curriculum. *Science and Education* 6:355–371.

Thier, H., and Nagle, B. (1994). Developing a model for issue-oriented science. In Solomon, J. and Aikenhead, G. (eds.), *STS Education: International Perspectives on Reform*, Teachers College Press, New York, pp. 75–83.

Waks, L. J. (1987). Afterword: The STS prophets and their challenge to STS education. *Bulletin of Science, Technology and Society* 7:1001–1008.

Welch, W. W. (1969). Curriculum evaluation. *Review of Educational Research* 39:429–443.

Yager, R. E. (1992a). Science-technology-society as reform. In Yager, R. E. (ed.), *The Status of STS: Reform Efforts around the World*, ICASE 1992 Yearbook. International Council of Associations for Science Education, Knapp Hill, South Harting, Petersfield, pp. 2–8.

Yager, R. E. (ed.) (1992b). *The Science, Technology, Society Movement*, National Science Teachers Association, Washington, D.C.

Yager, R. E. (ed.) (1996). *Science/Technology/Society as Reform in Science Education*, SUNY Press, Albany, NY.

Yager, R. E., and Krajcik, J. (1989). Success of students in a college physics course with and without experiencing a high school course. *Journal of Research in Science Teaching* 26:599–608.

Ziman, J. (1980). *Teaching and Learning about Science and Society*, Cambridge University Press, Cambridge.

Ziman, J. (1984). *An Introduction to Science Studies: The Philosophical and Social Aspects of Science and Technology*, Cambridge University Press, Cambridge.

CHAPTER 4

Trade-offs, Risks, and Regulations in Science and Technology
Implications for STS Education

Julie C. DeFalco

INTRODUCTION

Risk is inherent in life. Just by stepping out of bed in the morning, you set in motion a chain of events that can in some sense be risky. Taking a shower is hygienic and makes you pleasant to be around, but you could slip and hurt yourself in the tub. Having eggs for breakfast gives you energy, but they also contain cholesterol. You need to walk outside to go to the bus stop, but on the way, you could fall on the sidewalk. And if you are too frightened to leave the house, simply lying in bed incurs the risk of inactivity.

We accept these risks, and in doing so, often try to control them. We do this by calculating trade-offs: the "costs" and "benefits" of doing or not

Julie C. DeFalco, Competitive Enterprise Institute, Washington, D.C. 20036.

Science, Technology, and Society: A Sourcebook on Research and Practice, edited by Kumar and Chubin, Kluwer Academic / Plenum Publishers, New York, 2000.

doing a certain activity, or doing something else. As U.S. Supreme Court Justice Stephen Breyer (1993) notes,

> We find it worth spending money on an ordinary fire alarm system, but not worth installing state-of-the-art automatic-phone-dialing fire protection. We believe it worth installing guard rails on bridges, but not worth coating the Grand Canyon in soft plastic to catch those who might fall over the edge. (p. 16)

We could refrain from playing sports because we could be injured, but not exercising incurs serious health risks as well.

While trade-offs such as these are well known on the personal level, they also exist at all levels and in all activities. In physics, every action causes an equal and opposite reaction, both of which are clearly apparent. Unlike physics, however, when it comes to human actions, the consequences are frequently not seen. This is what the Nobel Prize-winning economist Milton Friedman meant when he famously said that there's no such thing as a free lunch. Someone had to grow the food, transport it, cook it, and serve it. Somehow, somebody has paid for that meal.

This chapter will discuss the sometimes invisible trade-offs which occur when policymakers attempt to limit risks in science and technology through laws and regulations. Federal, state, and local laws and regulations cover nearly every action we take: the manufacturing, purchasing, and marketing of products, the hiring and firing of labor, the attempts to protect the environment. Many of these laws have been passed because of a belief that our market economy has somehow failed—failed to protect consumers against ever-increasing applications of technology or faulty goods, failed to protect workers from hazardous conditions, or failed to save the environment from harm.

More often than not these government regulations "fail." They fail in ways more profound than the claimed market failure. Sometimes these laws demonstrably hurt those whom the laws intended to protect. Risk experts John Graham and Jonathan Wiener (1995) write:

> Americans are engaged in a national campaign to reduce risk. Yet confounding this national campaign to reduce risk is the phenomenon of "risk tradeoffs." Paradoxically, some of the most well-intentioned efforts to reduce identified risks can turn out to increase other risks. (p. 1)

One explicit example of this happened in Peru. Following the advice of the United States Environmental Protection Agency, one city in Peru stopped using chlorine to purify the municipal water supply because the use of this chemical may slightly increase the risk of some cancers. Because one of the most effective water treatments was banned, 3,000 people died in a cholera epidemic. Cholera is spread by contaminated water. After this tragedy, Peruvian officials repealed the chlorine ban (Anderson, 1991).

Sometimes measures to reduce risk actually increase it, but in a less obvious way. For example, the number one health risk is poverty. Justice Breyer (1993) writes that

> [Government] regulation of small risks can produce inconsistent results, for it can cause more harm to health than it prevents.... At all times regulation imposes costs that mean less real income available to individuals for alternative expenditure. That deprivation of real income itself has adverse health effects, in the form of poorer diet, more heart attacks, more suicides. (p. 23)

It is not just taxpayer dollars spent by the government on government-sponsored programs that decrease overall wealth. The amount of money businesses spend on compliance with federal regulations is greater than the amount the federal government spends on all domestic discretionary spending programs. It is triple the amount of the federal deficit and a third of the level of all federal outlays (including military and entitlement spending) (Crews, 1996). And because regulatory costs for businesses are passed on to customers, "regulatory costs exceed 19 percent of a household's after-tax budget" (Crews, 1996). Regulation is an invisible tax on the American people.

One way to look at how much the cost of regulation affects people is to look at what else people could do with this money. Economists' term for such an alternative expenditure is "opportunity cost"—the value of your next best option. We seek to minimize our opportunity costs in our daily lives. Frequently, when it comes to regulations, the opportunity cost of an action is not considered. This can lead to outcomes where we are worse off than before.

For example in the early 1990s, courts in New York City ordered the removal of asbestos insulation in schools because there is evidence that this material can cause cancer. Not only did this action raise the risk for the workers who removed the asbestos, and for the students in the schools (asbestos dust in the air is far more harmful than dormant asbestos slabs in the walls), but it also took the hundreds of millions of the city's dollars away from other municipal purposes, such as hospitals or police protection. Thus, not only were taxpayer dollars wasted, but people's health was more at risk than before. As Graham and Wiener (1995) point out:

> Unless policymakers consider the full set of outcomes associated with each effort to reduce risk, they will systematically invite such risk tradeoffs.... The net effect of actions taken to reduce risk is complex; the phenomenon of risk trade-offs suggests that in the national campaign to reduce risk, not as much health, safety, and environmental protection is being achieved as was intended and expected. (p. 2)

This chapter examines the effects of three laws and regulations: the Endangered Species Act, which intends to save certain flora and fauna from extinction; the Corporate Average Fuel Economy (CAFE) standards for passenger vehicles, which intends to cut down on oil imports; and the air bag mandate, which intends to increase car safety.

The track record of these rules shows that the risks they were supposed to ameliorate actually *increased* risk. Thus, that which was presented to the public as a "free lunch" turned out to be a poison pill.

STS AND THE ENDANGERED SPECIES ACT

Introduction

The Endangered Species Act (ESA) attempts to achieve a laudable goal: protect and conserve animals and plants in danger of extinction. Unfortunately, the ESA not only has failed to achieve its stated aim, but it has worsened the problem.

The ESA affords a good opportunity to explore many aspects of STS education. For example, a civics class could discuss how the political environment influences the creation of legislation, how a bill becomes a law, and how courts interpret laws that have been passed. A biology class could learn how biologists determine taxonomy. A math class could learn how statistics are used to determine populations of species.

Legislative History

Americans have always sought ways to prevent the extinction of animals. Since the 18th century, state and local laws have dealt with such conservation issues. By the early part of this century, private organizations had been instituted to monitor the populations of various creatures. The 1908 Lacey Act was the first federal law passed to deal with these issues. Using the interstate commerce clause of the Constitution, it forbade the interstate trafficking of wildlife if it broke a state law. In 1934, hunters and anglers lobbied states to impose taxes on certain sporting activities to fund conservation efforts, and this tax was soon expanded to cover the manufacturers of certain sporting equipment.

It was not until the 1960s, however, that major federal laws were proposed. The first of these laws, the Endangered Species Preservation Act of 1966, authorized federal government acquisition of land to provide habitat for wildlife at risk of extinction. This was only applicable to selected fish and wildlife species. It was not considered strong enough, so in 1969, the Endangered Species Conservation Act (ESCA) was passed. The ESCA not only covered species not native to the United States, but also barred trade in endangered species except in clear cases of economic hardship, and asserted coverage over reptiles, amphibians, mollusks, and crustaceans. It also required the Secretary of the Interior to set up an international meeting to deal with these issues. For that reason, in 1972, the Convention on International Trade in Endangered Species of Wild Flora and Fauna (CITES) was created. Congress passed today's ESA in 1973 as a way to comply with the international treaty, despite the fact that CITES had not yet been ratified. The ESA expanded the government's jurisdiction even further than the previous acts did, and it contained several provisions that have led to great controversy. Consequently, the act was amended in 1976, 1977, 1978, 1979, 1982, and 1988, and still has reform legislation pending.

Today's Endangered Species Act

The ESA sounds like a simple law. The federal government makes a list of animals or plants in danger of extinction, and then monitors and protects the creature until it is recovered, before finally removing it from the list. In practice, however, the law is extremely complicated. "The Endangered Species Act is 'the pit bull of environmental laws,' said Donald Barry (Adler, 1995), assistant secretary for Fish, Wildlife, and Parks at the Department of the Interior. 'It's short, compact, and has a hell of a set of teeth'" (p. 17)

Analysts Thomas Lambert and Robert Smith (1994) outlined three key provisions of the ESA:

- Section four of the ESA mandates that species be listed as endangered or threatened "solely on the basis of the best available scientific and commercial information regarding a species status, without reference to possible economic or other impacts of such determination";
- Section seven of the act, as it has been interpreted by the judiciary and implemented by the U. S. Fish and Wildlife Service, requires federal agencies to insure that the actions they authorize, fund, or carry out neither jeopardize the continued existence of a listed species nor modify habitat that is critical for its survival;
- Section nine states that no person may take an organism listed as endangered or threatened, where "take" means, "to harass, harm, pursue, hunt, shoot, wound, kill, trap, capture, collect, or to attempt to engage in any such conduct." The strength of this provision lies in the definitions of harass and harm. (pp. 6–8)

Problems with the Endangered Species Act

There are three main problems with the ESA:

- *The ESA infringes upon property rights.* While the ESA allows the federal government to restrict the use of privately owned land to provide habitat for a given species, nothing in the ESA requires the federal government to compensate the landowner. This goes against several hundred years of accepted common law and the American tradition of respecting private property. Even more important, the ESA violates the Fifth Amendment of the United States Constitution, which reads: "nor shall private property be taken for public use without just compensation."

- *The ESA is ineffective.* The purported goal of the ESA is to "recover" species. Although a total of 1,119 species have been listed since the law's inception,[1] only 27 have been removed from the list. Of these, seven became extinct while listed. Nine were officially delisted because of data errors (that is, new populations of the species were discovered after its listing or taxonomic errors were discovered). Only 11 were considered "recovered," although the evidence shows that even this is not true.

- *The ESA is unrealistic and expensive.* The ESA ostensibly requires that all species be saved at any cost. This is unrealistic because some species are biologically marginal and might become extinct anyway. Due to public pressure, the vast majority of ESA funds is skewed toward "popular" animals, such as whales and bald eagles, at the expense of less interesting creatures, such as insects. ESA alone costs the federal government hundreds of millions of dollars. This does not even count the costs the regulations impose on private landowners.

Property Rights

> When a species is listed, there is a freeze across all of its habitat for two to three years while we construct a habitat conservation plan which will later free up the land.
>
> Bruce Babbitt (1994), United States Secretary of Interior

---

[1] At the time of this writing, there were 890 endangered species listed in the United States (337 animals, 553 plants). There are 229 species listed as "threatened" (114 animals, 115 plants). Of the total figures presented, 451 are animals and 668 are plants.

The premise of the ESA is that wildlife belongs to all citizens in common. Therefore, government regulations and not private markets are necessary to ensure the survival of endangered species.

> The ESA is simply a manifestation of the public's interest in wildlife, and the public's aversion to paying for the satisfaction of that interest. Clearly, if "wildlife belongs to all citizens in common," then the presumption is that there is little reason to pay for something that "we" already own. Of course, it is unclear whether the public actually "owns" the wildlife.[2] Conversely, private property rights in land are well defined, if not well defended. (Sugg, 1993–1994, p. 11–12)

Contrary to the assertions of many ESA proponents, the law heavily infringes upon private property rights. As Secretary Babbitt's quote illustrates, the ESA's primary consideration is the habitat of the species and not whatever plans the landowner had for the property or how much a "freeze" on the property will cost the landowner.

A 1978 Supreme Court case, *Tennessee Valley Authority vs. Hill*, enshrined this principle when it declared "that 'the balance has been struck in favor of affording endangered species the highest of priorities' and that 'the plain intent of Congress in enacting this statute was to halt and reverse the trend toward species extinction, whatever the cost.'"(Sugg, 1993–1994, p. 11). Though the legislative history of the ESA—what Congress said when it wrote the law—does not support this contention, this case has been the basis for subsequent decisions against the rights of landowners.

If the public wants to have land set aside for endangered species, the public ought to pay for it instead of forcing a single party to finance it. When the federal government decides to build a highway or military base, the principle of "eminent domain" allows it to take the necessary land, but the Fifth Amendment guarantees that the landowner receives just monetary compensation (whether that compensation is accurate is another issue entirely). Though the ESA is not an overt act of eminent domain, it nonetheless, in practice, constitutes a "regulatory taking" of the land. When a piece of land falls under the ESA, the owner's ability to build on it or otherwise develop it can be prohibited. For that reason, its market value drops. People have lost their life savings in this manner.

The government has clearly overstepped the boundaries of the original intent of the legislation. Today, regulatory takings cover not only

---

[2] At least one federal court case plainly denies such ownership to the government (see *Hughes vs. Oklahoma*, 441 U.S. 322 (1979), in which the court expressly abolishes such state ownership).

the territory the species is known to inhabit, but also territory the species *might* use. There has even been an effort to set aside whole "ecosystems" for preservation. It is called the Habitat Conservation Plans. These plans proceed, notwithstanding that "ecosystems, cannot be objectively delineated. Discussing the development of the 'ecosystem' concept, renowned ecologist Paul Colinvaux wrote, 'The idea was that patches of earth, of any convenient size, could be defined and studied to see how life worked there.' In other words, ecosystems are units of nature arbitrarily defined by humans for their scientific convenience" (Lambert and Smith, 1994, p. 50). Even Secretary Babbitt has admitted this, saying that ecosystems "are in the eye of the beholder" (Lambert and Smith, 1994, p. 52).

The worst aspect of this abrogation of property rights is that it leads to perverse incentives. Since the ESA brings with it a host of financial and legal headaches, the property owner might decide to either destroy the species before the government discovers its presence or destroy its habitat before a species settles there. In the Pacific Northwest, this practice is known as "shoot, shovel, and shut up" (Vivoli, 1992). As Sam Hamilton, Fish and Wildlife Service administrator for the state of Texas noted, "The incentives [of the ESA] are wrong here. If I have a rare metal on my property, its value goes up. But if a rare bird occupies my land, its value disappears. We've got to turn it around to make the landowner want to have the bird on his property" (Seasholes, 1995, p. 8).

Even one prominent supporter of the ESA, Michael Bean, chair of the Environmental Defense Fund's Wildlife Program, acknowledged this in a seminar for FWS employees (Bean 1994):

> Despite nearly a quarter of a century of protection as an endangered species, the red-cockaded woodpecker is closer to extinction today than it was a quarter of a century ago when the protection began. There is, however, increasing evidence that at least some private landowners are actively managing their land so as to avoid potential endangered species problems. The problems they're trying to avoid are the problems stemming from the act's prohibition against taking endangered species by adverse modification of habitat. And they're trying to avoid those problems by avoiding having endangered species on their property.... Now it's important to recognize that all of these actions that landowners are either taking or threatening to take are not the result of malice toward the red-cockaded woodpecker, not the result of malice toward the environment. Rather they're fairly rational decisions motivated by a desire to avoid potentially significant economic constraints. In short, they're really nothing more than a predictable response to the familiar perverse incentives that sometimes accompany regulatory programs, not just the endangered species program but others. (no page number)

The ESA's infringement on property rights not only demonstrably hurts people, but it hurts endangered species as well.

Table 1. Partial List of Delisted Endangered Species

Common Name	Historic Range	Date Listed	Date Delisted	Official Reason (Real Reason)
Brown pelican	USA (AL, FL, GA, SC, NC, etc.)	10/13/1970	2/4/1985	Recovered (DDT Ban)
Arctic peregrine falcon	USA (AK), Canada, Greenland	10/13/1970	10/5/1994	Recovered (DDT Ban)
Palau dove	W. Pacific: USA (Palau islands)	6/2/1970	9/12/1985	Recovered (Data Error)
Palau fantail flycatcher	W. Pacific: USA (Palau islands)	6/2/1970	9/12/1985	Recovered (Data Error)
Palau owl	W. Pacific: USA (Palau islands)	6/2/1970	9/12/1985	Recovered (Data Error)
Rydberg milk-vetch	USA (UT)	4/26/1978	9/14/1989	Recovered (Data Error)
Gray whale	USA (CA, OR, WA, AK), Canada, Russia	12/2/1970	6/16/1994	Recovered (1937 Whaling ban)
American alligator	USA (AL, AR, FL, GA, LA, MS, NC, OK, SC, TX)	3/11/1967	partial delistings 1975–1985	Recovered (Data Error)
Red kangaroo	Australia	12/30/1974	3/9/1995	Recovered (Data Error)
Eastern gray kangaroo	Australia	12/30/1974	3/9/1995	Recovered (Data Error)
Western gray kangaroo	Australia	12/30/1974	3/9/1995	Recovered (Data Error)

Reprinted with permission from "Delisted and Endangered Species" Competitive Enterprise Institute, Washington, D.C., April, 1997.

Efficacy of the ESA

Despite impressions to the contrary, the Endangered Species Act has not saved a single species. Table 1 lists some of the ESA's "success stories." None of these recoveries had anything to do with the ESA (CEI, 1997):

- The main factor in the recovery of the Eastern brown pelican and the arctic peregrine falcon was the banning of the pesticide DDT, which caused reproductive failure in these birds. DDT was banned in 1972, one year prior to the enactment of the ESA. In the case of the brown pelican, reintroduction efforts by the Louisiana Department of Wildlife (independent of the U.S. FWS) contributed to its recovery. As for the arctic peregrine falcon, "it was the remoteness" of the nesting habitat in northern Alaska, according to Jay Sheppard, formerly of the FWS' Office of Endangered Species, that contributed to the bird's rebound, not actions taken under the ESA.

- The Palau dove, the Palau fantail flycatcher, the Palau owl, and the Rydberg milk-vetch are examples of recovery due to data error. A 1988 U.S. General Accounting Office report noted that "according to FWS officials ... the three Palau species owe their 'recovery' more to the discovery of additional birds than to successful recovery efforts" (GAO, 1998). As for the milk-vetch, a small plant, new specimens were discovered that brought its numbers from an estimated 2,000 to an estimated 200,000—a 10,000 percent increase. Said Jon L. England, the FWS botanist who wrote the final rule delisting the milk-vetch, "it was essentially a data error."

- The recovery of the gray whale had nothing to do with the ESA. Once prized for their useful oil, whales became less in demand when the market for oil collapsed in the early part of the 20th century (because of the development of cheaper substitutes). Whaling became viable again in the 1920s, but was banned in 1937 by an international treaty. The U.S. passed the Marine Mammal Protection Act in 1972, one year before the ESA, and, even more important, the Mexican government began to protect calving lagoons.

- The recovery of the alligator, the most famous example of the ESA, turned out to be due to the discovery of a data error. The National Wildlife Federation has even stated that "it now appears that the animal never should have been placed on the Endangered Species List; recent evidence suggests that the 'gator was thriving in some parts of its range throughout the 1960s, albeit at somewhat lower population levels than now exist." (Lewis, 1987) Indeed, it has been suggested that the ESA hampered the states of Florida and Louisiana, which tried to manage their alligators as a renewable resource and thus make protecting alligators profitable.

- The ESA had no effect at all on the recovery of the three kangaroos listed, for they are all native to Australia and the ESA has no jurisdiction there. Incidentally, these kangaroos were never in danger of extinction; their population fluctuates with rainfall changes, but they number in at least the hundreds of thousands. The FWS was simply looking for a foreign species to include on its list, and the kangaroo seemed appealing.

The act's supporters claim the ESA saves species when all the evidence is clearly to the contrary. The problem is not that people don't care about endangered species, but that the egregious regulatory regime engendered by the ESA, with its perverse incentives, guarantees the act will never work.

Cost of the ESA

Perhaps the Supreme Court was only using uplifting rhetoric in its 1978 decision to preserve endangered species at "whatever the cost." But this virtual mandate has turned out to be extremely expensive. Considering the ineffectiveness of the ESA, and its imposition on private landowners, this is highly disturbing.

As Lambert and Smith (1994) pointed out, the implementation cost of this act is very high. For example, the total reported spending in 1992 for endangered species protection is $290 million, and the inspector general of the Department of the Interior estimates $4.6 billion is needed to recover all presently known species. For every dollar FWS spends on recovery of endangered species, it spends another $2.26 on consultation, permitting, law enforcement, and research done to list species.

High as they are, these figures also do not take into account the costs to landowners who must comply with these regulations and whose land often loses real value because of the ESA's strictures. Given the poor record of the ESA, this money is wasted. The opportunity cost of these resources—what else could have been done with those funds—is very high.

As long as the government does not have to pay for the land it takes by regulation, it can avoid making difficult decisions about what is most important. We all have to work within budget constraints (most of us drive Hondas instead of Rolls Royces) and there is no reason why the government should not have similar limits. The only way to do this is to require government to pay for the land it wants to reserve for endangered species.

Counting money wasted on a dilatory regulation is one thing. It is quite another to determine what the social costs of the regulation are. Here are two examples of how the ESA has demonstrably hurt people:

The Stephens kangaroo rat and the California fires. To minimize the damage to people and property, state and local regulations in Southern California require landowners to remove flammable vegetation around structures because of the fires that annually sweep through the area. The best method of clearing this brush away is a practice called "disking," a process by which the top layer of soil is overturned, burying the vegetation safely beneath the ground.

In 1989, these state and local regulations came into direct conflict with the federal ESA. The Fish and Wildlife Service ordered the people of Riverside County, CA to mow (instead of disk) their property in order to preserve the habitat of the endangered Stephens kangaroo rat (k-rat). Mowing will not disturb the k-rats' burrows, but neither does it clear the vegetation away.

In October 1993, a fire swept through Riverside County, burning 25,000 acres and destroying 29 of 300 homes in its path. Nineteen of these homes were in designated k-rat "preserve study areas" and for that reason, the homes' owners were not permitted, under the ESA, to properly protect their property by disking.

One couple, Andy and Cindy Domenigoni, had let 800 of their 3,200 acres of farmland lie fallow. Into the undisturbed land came the k-rat, and soon after, ESA restrictions. Though local regulations required them to remove the underbrush on the property, the federal government forbade disking, the only effective way of clearing all of it. They did manage to clear a small area—seven acres. This field was where the Domenigonis, with their 100 head of cattle, waited out the fire which consumed the rest of their property. In the aftermath of the fire, it was discovered that the k-rat was no longer in the area—not because of the fire, but because the brush and weeds had grown too thick (Lambert and Smith, 1994).

The Domenigonis' neighbor, Michael Rowe, fared slightly better on that night. As he saw the fire approaching, he went out with a tractor in the middle of the night and disked a firebreak between his house and the Domenigoni field, even though disking was still technically forbidden. Despite an unfortunate shift in the wind which blew the fire toward his land, Michael Rowe saved his home.

"There's an inherent conflict between preserving wildlife and fire safety," said Richard Wilson, director of the California Department of Forestry and Fire Protection (Sugg, 1994, p. 4). In this case, the k-rat received more consideration than did the people who owned its habitat.

The Golden-Cheeked Warbler. Travis County, Texas contains some of the best land in the state, and, in 1990, discovered it had one of the worst problems: an endangered neotropical migratory songbird called the golden-cheeked warbler.

The FWS ordered so much private land to be set aside for the warbler that "according to the county's deputy chief tax appraiser, county property values decreased nearly $360 million . . . Before the birds were added to the endangered-species list, the market appraisal of this real estate was ten times greater than after their listing" (Lambert and Smith, 1994). In one case, a woman who had purchased 15 acres as an investment to supplement her retirement income saw the value of her land plummet from $830,960 in 1991 to $30,380 in 1992 as a direct result of the development prohibitions of the ESA (Kazman, 1995).

The case of the golden-cheeked warbler also shows that the FWS has overstepped its bounds. The FWS not only forbids development on land that is demonstrably warbler territory, but it also freezes development on nearby land. "No actual members of a listed species needed to be present for FWS to preclude someone from using their own land under its 'harm regulation.' [As the FWS has written,] 'Although [the] development area does not contain occupied warbler habitat. . . . The service currently believes that development activities in general will cause indirect impacts to the warbler. . . .'"(Kazman, 1995, pp. 4–5).

Because people felt threatened by the ESA regulations, they did their best to avoid encountering these birds in the first place. "While I have no hard evidence to prove it, I am convinced that more habitat for the black-capped vireo [another endangered bird] and especially for the golden-cheeked warbler, has been lost in those areas of Texas since the listing of these birds than would have been lost without the ESA at all," said Larry McKinney, director of Resource Protection at the Texas Parks and Wildlife Department (Seasholes, 1995, p. 8).

Conclusion

An STS curriculum studying this perspective of the ESA would benefit students and better prepare them to evaluate trade-offs in public policy. It neatly captures the perils of imposing a law without evaluating its consequences. We all see value in a diversity of wildlife. However, conserving species is only one of many competing goals of society—a fact sometimes lost in the course of studying other STS subjects. Society's other goals include preserving individual freedoms and respecting the Constitution. Resources consumed by ESA mandates are resources unavailable for other projects. This point is especially pertinent because it is overwhelmingly clear that the ESA, after 25 years, cannot claim a single recovery.

There are other, better ways to promote conservation while retaining our individual rights. First, FWS' ability to prohibit land use must be eliminated, and in the event that land is taken, property owners must be compensated. Otherwise, the perverse incentives will continue.

   Second, we must learn from the efforts made by private individuals and groups long before the ESA was ever conceived. Government regulation tends to be arbitrary and restrictive. In contrast, in a market economy, people can be free to express their values and can experiment with innovative, creative ways of preserving species. One environmental group has claimed that "when a man puts a price tag on a wild animal, that wild animal eventually disappears" (Struzik, 1992, p. 24). Indeed, in practice, we have seen quite the opposite. Harnessing the profit motive and self-interest has repeatedly proven successful for both human and beast.

   This can mean buying land and keeping it in trust simply to provide land for wildlife. A good example of this is the Nature Conservancy (TNC). TNC operates a key migratory bird corridor in Cape May, New Jersey. Another way is through private conservation funds, which have been instrumental in making the wood duck, as well as all three species of American bluebird, common once again (Seasholes, 1995). Numerous other organizations exist to protect other animals, such as the Peregrine Fund (for peregrine falcons). Companies such as International Paper have implemented innovative forest management strategies to preserve a variety of wildlife. They also charge admittance fees that are used for these conservation efforts. Internationally, in Zimbabwe, local farmers have benefited from "privatization" of wildlife such as elephants, and in Papua, New Guinea, profits from butterfly farming have encouraged locals to refrain from cutting down valuable forest land.

   America has a long tradition of private conservation. Government regulations of the past 30 years have stifled this innovation, replacing it with an inflexible bureaucracy. Market solutions are dispersed, enabling creative solutions and approaches tailored to specific, local conditions. Instead of wasting resources lobbying the government to act in one way or another, a better use of this time and money would be toward protecting species individually. We should revive this venerable tradition once again as the best way of ensuring the survival of species.

STS AND CORPORATE AVERAGE FUEL ECONOMY STANDARDS

Introduction

   Passed during the oil crises of the 1970s, the federal government's Corporate Average Fuel Economy (CAFE) standards were specifically intended to reduce America's dependence upon foreign oil. More than 20 years later, CAFE is now being promoted as a way to save the environment. While the evidence for CAFE's effectiveness in achieving either of these goals is slim to nonexistent, it is demonstrably true that CAFE has had a serious, unintended effect: By causing vehicles to be smaller and lighter than they otherwise would be, CAFE kills people.

CAFE is a good issue to include in STS education because it clearly illustrates how even well intended ideas and good science can result in laws that have unintended, negative effects. A history class might study the political environment leading up to CAFE's passage in 1975, including America's relations with the Middle East, and the rise of OPEC (Organization of Petroleum Exporting Countries). An economics class might study the effects of price-controls and rationing; artificially low prices for a product means that suppliers may not be willing to produce much of it, and artificially high prices mean there might be less demand for it. A chemistry class might study how auto emissions are measured and examine the effects of the mix of emissions.

Legislative History

Price controls and rationing of many goods were initially implemented during the Nixon administration and continued throughout the Ford and Carter administration. These policies exacerbated, rather than ameliorated, the economic problems throughout the 1970s. The OPEC oil embargo of 1974 was the major crisis. Gas prices shot out of control. Commentators of the period predicted that gas prices would skyrocket by the mid-1980s, perhaps reaching $2.50/gallon (in 1975 dollars). The resulting recession led politicians to seek legislative solutions to quell the public's ire. Because the price controls and rationing had failed, politicians sought another way to counteract the effect of the oil embargo.

One way to handle this problem, many believed, was to decrease the United States' dependence upon foreign oil. Accordingly, CAFE standards were made into law in 1975. Under CAFE, each auto manufacturer selling in the U.S. was required to make its entire passenger car fleet meet a specified fuel economy target every year. CAFE was gradually increased over the years. Currently, for passenger cars, the CAFE standard is 27.5 miles per gallon, and for light trucks and vans, the standard is 20.7 miles per gallon. Manufacturers who failed to meet CAFE standards would be fined. Congress charged the National Highway Traffic Safety Administration (NHTSA) with administering the program and monitoring compliance.

CAFE is always a work in progress. Oil prices were so low in the 1980s that more consumers bought large cars. As a result, auto companies requested that CAFE be relaxed. NHTSA decreased it for model years

1986–1989—four years in all—before increasing it again in 1990. Congress picked up on this trend, and many members began a strong push to increase CAFE, resulting in numerous bills. The most important was that supported by Senator Richard Bryan (D–Nevada), which would have raised CAFE to 40 miles per gallon by the year 2000. "Despite opposition by the Bush administration and automakers, the Bryan bill came close to passing the Congress in both 1990 and 1992" (Graham and Wiener, 1995, p. 89). Raising CAFE was even suggested in Bill Clinton's presidential campaign manifesto, "Putting People First" (Clinton, 1992).

Meanwhile, the rationale for CAFE changed. Concerns about America's dependence upon foreign oil gave way to concerns about the amount of vehicle emissions. In the 1990s, there have been many proposals to increase CAFE because of its supposed environmental benefits. To date, all have been successfully opposed. However, the signing of the Kyoto greenhouse protocols in 1997, which, if implemented, would commit the U.S. to reduce carbon dioxide emissions, have given pro-CAFE forces another opportunity to push for an increase in CAFE. This would have a negligible impact on the environment, but a disastrous impact upon auto safety.

Corporate Average Fuel Economy Standards and Oil Imports

Despite the dire predictions of the 1970s, gas prices actually plummeted after 1982 when President Reagan lifted the price controls on gasoline. In 1995, gas prices reached their lowest in history. Foreign oil is cheaper to produce than American oil. That means when gas prices are low—as they have been in recent years—American oil is generally too expensive to be competitive, and thus, more gasoline is imported. As a result, "oil imports have risen from 35 percent in 1974 to more than 50 percent in 1995" ("Minivans," 1997).

CAFE was initially proposed on grounds of national security. Presumably America would be less dependent upon oil and therefore less affected by oil price changes. But CAFE doesn't affect only foreign oil sales—it affects *all* oil sales. Thus, we would also see less oil production in the U.S. as well. If the overall reduced demand for oil still depends upon imports from volatile areas, then CAFE hasn't done anything to forward national security.

It is important to note that mandating decreased oil imports on national security grounds is highly debatable. One could argue that it makes more sense to husband America's oil until an emergency, when it would be available as a backup, rather than force people to use the relatively more expensive American oil today.

Corporate Average Fuel Economy Standards and the Environment

Today, much public debate is centered not on decreasing America's dependence upon foreign oil, but on finding solutions to environmental problems, particularly pollution. Along with numerous federal, state, and local regulations determining the quantity of certain emissions permitted in the air, CAFE has been portrayed as a way to help reduce these emissions—some of which have been accused of contributing to global climate change (Since this is a contentious related issue, we will simply say that the effect of these emissions on the climate is debatable). The Sierra Club has claimed that an increase in CAFE is "the biggest single step to curbing global warming" ("Single Step," undated).

Total emissions may not fall with a higher CAFE for several reasons. Cars and light trucks make up only 1.5 percent of all global human-made greenhouse gas emissions. Hiking CAFE standards by 40 percent would only reduce those emissions by 0.04 percent, according to the government's own data (DeFalco, 1997).

All new cars must meet the same emissions standard (measured in grams per mile) regardless of the car's fuel efficiency. Thus, the emissions—known as "volatile organic compounds, or VOCs—from all cars and light trucks must fall under the same ceiling. "Manufacturers therefore design engines and emission control systems to just meet the grams per mile standard, so improved fuel efficiency does not necessarily lead to fewer VOC emissions. More fuel efficiency is just as likely to result in less stringent emission control systems that still meet the grams per mile standard" (Graham and Wiener, 1995, p. 93). Though it may be likely that a larger vehicle would emit more than a smaller vehicle, it is possible that the opposite may be the case. Just as a person running faster than another person would likely perspire more, a smaller vehicle's engine must work harder and may therefore emit more pollutants.

Because confusing regulations could arbitrarily change, and because pollution control equipment often adds weight to a car, the National Research Council wrote in a 1992 report that "compliance with [emissions] standards will make it difficult to introduce more fuel efficient vehicles" (NRC, 1992, p. 75).

Despite their overall support for increased CAFE, the federal government's Office of Technology Assessment and the National Academy of Sciences have also suggested that CAFE might make people drive more, thwarting CAFE's entire purpose. In the words of Graham and Wiener (1995):

> By reducing the amount of fuel needed to drive a vehicle each mile, improved
> fuel efficiency lowers the driver's cost of operating a vehicle per mile of travel,

and thereby encourages drivers to travel more miles. More total miles of driving thus offsets some of the energy conservation gains achieved by the lower fuel use per mile of more fuel-efficient vehicles. The magnitude of this offsetting effect is uncertain, but it is estimated that somewhere between 10 and 30 percent of the potential fuel savings from increased CAFE standards will be lost to increased driving...." (p. 90)

Any benefits from CAFE would be delayed because the rule retards the sale of new cars. CAFE requires manufacturers to make smaller cars. If people would rather buy larger cars, then manufacturers must subsidize the price of small cars (to increase demand) or raise the price of larger cars (to decrease demand) in order to achieve CAFE goals and avoid fines. This leads to several situations: consumers holding onto their older, more polluting cars for longer periods of time; and, recently, people buying light trucks and minivans instead of large cars. This is what led to the slight decrease in CAFE in the mid to late 1980s.

Corporate Average Fuel Economy Standards and Auto Safety

Decades of auto research have shown that a large car is almost always safer than a small car. "Traffic safety analysts have found that occupants of lighter cars incur an elevated risk of serious injury and death in crashes compared to occupants of heavier cars. This statistical association has been demonstrated for both single-vehicle and multivehicle crashes ... *The negative relationship between weight and occupant fatality risk is one of the most secure findings in the literature*" (emphasis added) (Crandall and Graham, 1989, p. 110). A large car with airbags is still safer than a small car with air bags. In a two-car collision, occupants of the larger car come out better than the occupants of the smaller vehicle. In a collision with a tree or a wall, large cars are safer, because they have more "crush space" than do small cars.

This simple physical fact is often overlooked yet it is crucial to understanding the deadly effects of CAFE. "Most of the improvements in new car fuel economy were achieved before 1985" (Graham and Weiner, 1995, p. 89) by introducing new technologies, such as fuel-injected engines, and decreasing the size and weight of the vehicles.

Between 1979 and 1989, the average weight of a passenger car dropped by 1,000 pounds (roughly a 25 percent total reduction), and about half of this downsizing was due to CAFE. "The 500-pound ... reduction in the average weight of 1989 cars caused by CAFE is associated with a 14–27 percent increase in occupant fatality risk" (Crandall and Graham, 1989). By using updated fatality figures from the Insurance Institute of Highway Safety, it is estimated that in 1996, 2,700–4,700 passenger car occupant deaths (of a national total of 22,000) were the result of CAFE's downsizing effect (DeFalco, 1997).

One federal judge (*CEI vs. NHTSA*, 1992) criticized the increase in CAFE after model year 1989 in a 1992 case on this issue:

> Choice means giving something up. In deciding whether to relax the previously established CAFE standard for 1990, NHTSA confronted a record suggesting that refusal to do so would enact some penalty in auto safety. Rather than affirmatively choosing extra energy savings over extra safety, however, NHTSA obscured the safety problem and thus its need to choose. . . . [Instead, NHTSA] fudged the analysis, held the standard at 27.5 mpg, and, with the help of statistical legerdemain, made conclusory assertions that its decision had no safety cost at all. That is what it chose." (pp. 322–324)

Unfortunately, many of those who favor increases in CAFE de-emphasize or dismiss this trade-off. "We can safely raise CAFE standards while saving oil, lives and dollars," wrote Sierra Club activists Ann Mesnikoff and Steve Pedery (Mesnikoff and Pedery, 1996, no page number) in a statement typical of CAFE's supporters. They state traffic fatalities have declined since CAFE's enactment, that a wide assortment of automobiles are available to the American public, and some small cars fare better in crashes than do sports utility vehicles (SUVs). These statements are misleading.

It is true traffic fatalities have declined since 1975. But a broader picture is necessary to completely show the whole situation. Traffic fatalities have been steadily declining since the 1920s. This still does not answer the question of what would have happened in a world without CAFE. The available evidence—the thousands of fatalities attributable to CAFE—suggests those figures would have declined even further had CAFE never existed.

That there are many sizes of vehicles available to the public is an equally irrelevant argument. The issue is that these vehicles have all been affected by CAFE downsizing. Consumers have repeatedly demonstrated that they are not interested in buying small cars. Every year the Environmental Protection Agency puts out a list of the top ten most fuel efficient cars. These ten cars are never the same as the top ten most purchased cars. The proliferation of SUVs on the road clearly demonstrates that consumers seek size, power, space, and safety over fuel efficiency in the automobiles they choose.

It is also true that some SUVs may fare as badly as small cars do in single car crashes. This is because the center of gravity on many SUVs is higher than it is in passenger cars, making SUVs somewhat more unstable. But this argument does not address the fundamental issue of whether large passenger cars are safer than small passenger cars. It is quite clear that increasing the mass of a vehicle reduces the risk its occupants face in a crash.

The safety of SUVs is an important question. The most recent criticism leveled against SUVs is that they are more dangerous to occupants of other cars when an SUV and a passenger car crash. This has led to a call for an increase in CAFE for SUVs, instead of a decrease in the CAFE for passenger cars.

NHTSA issued a report in summer 1997 that stated there would be 40 fewer fatalities annually if the weight of SUVs were reduced by 100 pounds (NHTSA, 1997). Although the report specifically said this figure was not statistically significant (i.e., it was likely the result was by chance), this fact was heavily promoted in news reports about the study. Consequently, this calculation has been put forth as evidence that the fuel economy standard for SUVs should be reduced.

Unfortunately, the same press coverage omitted a much more important point. The same NHTSA study said that by increasing the average weight of passenger cars by 100 pounds, more than 300 lives would be saved every year—a fact that *is* statistically significant. It is important to note the study also admitted a reduction in the size and weight of trucks made these vehicles "become less crashworthy" and lessened the amount of damage they caused to other vehicles. But all crashes are not multivehicle crashes. Walls and trees have not been downsized accordingly and can cause much damage and death. People rightly buy cars to protect themselves and their families, not the unknown passengers of other cars. Similar information should be openly discussed and debated in STS classrooms. The trade-offs of CAFE are not immediately apparent, but they are serious and require significant debate.

Conclusion

When the government takes steps to achieve a particular goal, its actions should be as straightforward and narrowly tailored as possible. Under this criterion, CAFE clearly fails. The onus is upon those who would push for downsizing cars—against the clear wishes of consumers—to show the benefit of the policy. It is clear there are no net benefits from CAFE. CAFE has not reduced fuel consumption nor has it reduced the necessity for oil imports. Yet, through its auto downsizing, CAFE has engendered tragic consequences.

If reducing fuel consumption is an important and worthy goal, then there are better and more honest ways of doing this, such as increasing taxes on gasoline. This would be the most direct way of achieving this goal. The more fundamental issue, however, is whether fuel consumption ought to be reduced. People use gasoline because they find it useful. Restricting its use means denying them a resource that is valuable in their lives. People ought to be able to buy as much gas as they are willing to pay for, and if that means they would like to buy a car with lower fuel efficiency, that is their decision.

It is hard to reduce the "costs" incurred by fuel consumption without reducing the benefits of fuel consumption as well. One important benefit of auto use is increased mobility, something that has benefited the less well off in our society. In a discussion of the often-intangible benefits of this "auto-mobility," philosophy professor Loren Lomasky writes, "Previously one either lived in direct proximity to one's work or else on a commuter rail line. . . . The coming of the motor car augmented the bargaining power enjoyed by workers. . . . Widespread automobile ownership meant that the geographical radius of possible employment venues was dramatically extended" (Lomasky, 1995, pp. 11–12). No longer were people tied to a specific area so that they could be near their jobs. Dr. Lomasky (1995) contends "the automobile is, arguably, rivaled only by the printing press and the microchip as an autonomy-enhancing contrivance of technology" (p. 10).

As well as the goals stated previously, an STS class may also use CAFE as a way to study how a law supported by divergent parties (in CAFE's case, the legislation currently has the support of the federal government, consumer groups, environmental groups, and the auto industry) becomes virtually impossible to have a candid discussion about the trade-offs. While many groups not only defend CAFE, but also argue that it should be made more stringent, not a single one has forthrightly dealt with its safety consequences. In the case of CAFE, these various parties find it difficult, for reasons of money or political reputation, to admit CAFE is a serious problem. This presents an interesting angle which may be included in an STS class: why certain groups are tied to policies in the face of clear evidence to the contrary.

In the end, CAFE is primarily a political bill with dangerous consequences. There is very little hard scientific evidence to support CAFE; indeed, the evidence available suggests CAFE should be repealed completely.

STS AND AIR BAGS

Introduction

Like CAFE and the ESA, air bags were initially promulgated in the 1970s, a period in which an active government asserted its wisdom over that of the average citizen. Air bags were conceived as a way to provide "passive protection" to vehicle occupants in a day when most people did not wear seatbelts. They were introduced gradually at first, but were eventually mandated. It was not until reports of children, small women, and elderly people being severely injured and, at times, killed by air bags that many people began to think more critically of air bags.

Like other topics discussed in this chapter, air bags illustrate how some political forces (in this case, the federal government, so-called auto safety groups, and the auto industry) can band together and attempt to frame and reframe how the public views air bags. Air bags also can be used to study physics: how the force of the activated air bag counteracts the force of a crash, and how this can sometimes save lives, and sometimes be deadly for an occupant.

Legislative History

Air bags were originally developed in the 1960s as a passive restraint to protect the vast majority of drivers who were not accustomed to wearing seatbelts, an active restraint. As Brian O'Neill (1997), president of the Insurance Institute for Highway Safety, a nonprofit group funded by the auto insurance industry writes, in the latter part of the decade, lap belt use was 20 percent, shoulder belt use was as low as 2–5 percent, and lap–shoulder belt use was not even required until 1973. It should be noted that the purpose of air bags was to protect people in frontal crashes.

In 1977, NHTSA ruled that by 1982, new cars had to have passive protection (either air bags or automatic seatbelts) for occupants. This rule was challenged and revised, coming out in final form in 1984. That rule required a phase-in of either of these passive restraints, starting with 1987 model vehicles. In 1991, Congress declared that by September 1996, 95 percent of passenger cars must have air bags (to be increased to 100 percent by September 1997). A similar rule was passed for light trucks and vans. NHTSA was charged with administering this mandate. People with special medical problems could get permission to deactivate their air bags, but this was difficult to obtain, and it was difficult to find a dealer or mechanic who would deactivate the air bag.

Due to public concern and outcry about the unintended effects of air bags after the mandate was fully in force, in November 1997, NHTSA released a compromise rule. The compromise said that air bags would still be mandatory equipment in cars, and people still would not be permitted to turn them off at will. However, beginning in January 1998, citizens falling into one of four risk categories could fill out a form and receive permission from NHTSA to have an on–off switch attached to one's air bag. (These risk categories are: unavoidably using a rear-facing infant seat in the front passenger seat; unavoidably carrying children in that seat; sitting closer than 10 inches to the steering wheel; or having a medical condition putting one at risk.) Although the agency has said it would not check up on the verity of people's claims, those who lie on these forms could conceivably receive up to five years in federal prison.

The Efficacy of Air Bags

Air bags have not performed as well as originally predicted, but they are still credited with saving 1100–1600 lives through part of 1996 (Claybrook, 1997). There are many ways to evaluate how effective airbags are.

Air bags are installed in the steering column or on the dashboard on the passenger side (some manufacturers also offer side impact air bags in certain models, although these are not required by law). A sudden decrease in speed (as in a crash) prompts the air bag mechanism. In less than a second, a combination of chemicals is ignited, causing the bag to be filled with an expanding gas (pyrotechnical inflation). The bag bursts forth at approximately 200 miles per hour, preventing the driver from hitting the windshield and steering wheel.

The air bag's force alone has killed or injured more than 50 people, a problem to which we shall return. In addition, the loud noise made when the air bag expands has reportedly caused hearing problems, and many people have incurred slight or major burns from the chemicals used to ignite air bags. One of the biggest complaints is that air bags are often set off in relatively minor collisions, such as parking lot fender-benders, serving no protective function. Nonetheless, it is demonstrably true that many people's lives have been saved by air bags.

Mandating Air Bags

The original claims for air bags were extensive. First, air bags were supposed to be the vanguard of passive protection. A 1977 NHTSA brochure for the general public described airbags as "systems that protect automobile occupants from collision injuries automatically, without the need to fasten belts or to take other actions" (Kazman, 1983, no page number). Another NHTSA brochure (1977) issued during that time asked, "Is it necessary to wear a lap belt with an air bag?" The answer was no.

But there had already been some evidence that air bags were not only ineffective, they were dangerous. Sam Kazman, general counsel of the

Competitive Enterprise Institute in Washington D. C., who unsuccessfully sued NHTSA in 1979 over the air bag mandate, cited one study, entitled "Comparative Restraint System Evaluation Using Dummies and Cadavers in Car-to-Car Crash Tests," by Thomas Glenn of NHTSA's Office of Vehicle Safety Research. Kazman (1996) said this study, which showed how air bags might be less effective than lap–shoulder belts in certain types of crashes, was kept from the public, and the author of the study was "threatened with disciplinary action." Also, it was not until NHTSA was threatened with courtaction did they release their study of several crash-related deaths in air bag-equipped cars that showed a higher fatality rate than conventional cars. For over two years, this study had been kept from the public.

Supposedly, air bags had been extensively tested. According to Joan Claybrook (1997), administrator of NHTSA from 1977–1981, the air bag has undergone about 5,000 crash tests and computer tests since 1969, making it perhaps the "most tested" technology used in cars. Mrs. Claybrook had been responsible for the 1977 air bag rule and has since been a great supporter of them.

However, her agency admitted at the time that though there had been laboratory tests and simulations, there was no significant real world data to support the initial mandate. NHTSA noted this in 1977, claiming it could "force technology." NHTSA dismissed concerns thusly:

> It has been argued that the Department [of Transportation, of which NHTSA is a part] should not issue a passive restraint standard in the absence of statistically significant real-world data which confirm its estimates of effectiveness. Statistical 'proof' is certainly desirable in decision-making, but is often not available. (42 Federal Register 34,292).

Finally, air bags were supposed to be suitable for children. In 1977, NHTSA declared that the newer, pyrotechnically inflated air bags (the kind used today) would allow out-of-position children to be "pushed more gently out of the way" (42 Federal Register 34,292). As late as 1983, after she had left the government and became a consumer advocate, Mrs. Claybrook stated on television that air bags "fit all different sizes and types of people, from little children up to . . . very large males" (CNN, 1983, no page number).

Yet auto manufacturers had been aware of the problem since at least 1969, and had brought these concerns to NHTSA. These concerns were dismissed. In 1979, the General Accounting Office pointed out that NHTSA "does not share our concern about the potential injuries to out-of-position occupants" such as children. A 1979 Associated Press story quoted Mrs.

Claybrook as charging that General Motors delayed installation of air bags because of "a 'hurry-up style' finding that they hurt small children." Indeed, industry observers in 1979 pointed out:

> Both GM and Ford, which planned to offer air bags as options on some big cars in 1981, are concerned about a new problem: children. It seems the small fry are especially vulnerable because they're often sitting or standing in a position where they won't be properly restrained by air bags. Both companies feel that they could be subject to a myriad of product-liability suits involving children. Even if their air bag systems meet the letter of the law. (*Car and Driver*, 1997, p. 16)

Air Bag Debates

During the 1980s, air bags grew in popularity. To an extent, this enthusiasm was justified. However, as more and more cars, equipped with air bags, went on the road, the potential danger of these devices became more apparent to the public at large. In December 1992, NHTSA requested public comment on what information should be contained in permanent air bag warnings on vehicle sun visors. The American Automobile Manufacturers Association (an organization which represented the major car manufacturers until 1999) (AAMA, 1993) petitioned NHTSA to include explicit air bag warnings such as:

- "Air bags inflate with great force, faster than you can blink your eyes. If you are too close to the inflating air bag, it could seriously injure you."
- "An occupant who is too close to the inflating air bag can be seriously injured."
- "An inflating passenger air bag can seriously injure a child in a rear-facing child restraint" (American Automobile Manufacturers Association, 1993).

On the other hand, many self-named auto safety and consumer groups took the opposite stance, leading NHTSA to adopt a watered-down warning. These organizations told NHTSA:

- "Advocates [of Highway and Auto Safety] (1993) does not believe that a special notice or label regarding the means of obtaining the maximum protection from the . . . air bag needs to be permanently affixed to the vehicle. The required statement of such information in the owner's manual is sufficient."

- "It would be counterproductive to present this information [about the risks of airbags] by way of unnecessarily alarming statements as are proposed in this rulemaking" (AHAS, 1993).
- "If the NHTSA determines that [a label with a straightforward 'dos and don'ts' of airbags] is needed, no other informational statements, directions, or warnings should be permitted to confuse the intended message" (AHAS, 1993).
- "The Coalition [for Consumer Health and Safety] (1993) is concerned that motor vehicle manufactures may put their own language on labels that will inadvertently alarm motor vehicle occupants. . . . [We oppose] expanding upon the information proposed."
- "The warning not to sit too close or lean over the air bag implies that being too close to the steering wheel or the dashboard is a problem only in cars with air bags" (IIHS, 1993).
- "The proposed warning would mislead the public by implying that air bags can cause fatal or serious injuries that would not have occurred in a comparable vehicle without an air bag" (IIHS, 1993).

This is a strange position for organizations, who speak in the name of consumers and auto safety, to take. However, many of these groups had supported the air bag mandate all along (Joan Claybrook is a cochair of Advocates for Highway and Auto Safety), and it is not surprising they would continue to support the mandate despite the growing body of evidence against it.

By 1996, disturbing headlines drew national attention to the downside of air bags. Several people had been literally killed by their air bags. In one case, an infant was decapitated by an air bag hitting his rear-facing infant seat, which was in the front passenger seat. About 50 people, mostly elderly people, short women, and children, had been killed in this manner. Hearings held by the United States Senate and by NHTSA focused on the details of such accidents, and how the technology might be improved, but no one asked whether air bags ought to be a mandatory piece of equipment at all. (Senate Commerce Committee, 1997).

Conclusion

Although airbags are good in general, they are bad for children, short women, and the elderly. Ironically, these people have traditionally fallen into the groups of people society believes need extra protection. The question is not whether air bags should be in cars at all—many consumers consider airbags highly desirable. Rather, we should ask: Why must everyone must be forced to purchase air bags, even when some people intend to turn them off?

Although the governmental paternalism inherent in all such safety regulations—the idea that coercion is necessary "for our own good"—is troubling in itself, that is not the only reason to oppose the air bag mandate. Arguably, mandating air bags in autos has probably saved more people than if the air bags were simply optional equipment. But the mandate explicitly puts a specific, readily identifiable group at great danger.

Philosophy professor Loren Lomasky (1997) compares mandating air bags with mandating a vaccine: Even though the polio vaccine has mostly eliminated polio, each year a tiny percentage of people get polio from the vaccine itself; yet polio vaccination is mandated in this country. That is because it is impossible to predict who does and who does not need the vaccine, and therefore, it is a good bet for everyone to take the vaccine. Lemansky says that the analogy breaks down in the case of air bags because we know beforehand that children are more at risk than adults to get injured by air bags in car crashes. Therefore, the air bag mandate is an unconscionable public policy position.

Since ultimately the public has to buy the air bags (which add approximately $600 to the price of a new car), the public ought to decide for itself what it wants. One 1997 poll (IIHS, 1997) found 79 percent of respondents would want at least a driver-side air bag in their next vehicle, and 81 percent would feel safer in vehicles with air bags (although 68 percent thought children are more at risk in such vehicles). Another poll (CEI, 1997) showed with a ratio of 3:1, the public favors giving people the choice of purchasing a new car with or without an air bag.

Air bags were supposed to work for everybody and were supposed to work without seatbelts. Today, public officials and consumer groups alike claim air bags were never supposed to be used for children, and that they were always supposed to be used in tandem with seatbelts. Air bags were largely untested when they were initially mandated, and new findings about them were continually covered up. The most important lesson an STS class—and the public in general—can learn from the air bag story is that rules passed in the name of public safety can backfire. We, the public, cannot always expect groups and individuals, which have staked reputations upon the success of the law, to acknowledge that the original rule was a bad idea. This theme ought to be fully explored in STS classes, for it is especially egregious when these groups and individuals claim to be speaking in the name of "consumers," "the public," or "safety."

IMPLICATIONS FOR STS EDUCATION

One of the goals of STS education is to pave the way for a scientifically literate citizenry (Kumar and Berlin, 1996). Other goals include presenting an objective view of a science and technology-dominant world to students, using science and technology issues to engage students in critically analyzing related issues impacting society. From this viewpoint, it is necessary that STS education deal with the positive and the negative aspects of how science and technology interact with society.

Frequently, in STS, the discussion skews toward more regulation of how science and technology are used. Students should be exposed to the frequently unseen negative impacts of regulation. Teachers should also resolve to present both sides of the story, that is, enable a discussion of the human costs of regulations. STS education should challenge students to explore both sides of the issue, and curricula should reflect this.

When students are made aware of the pitfalls of government regulation they become better informed, which enables them, when they grow older, to become productive members of society. They will be better able to evaluate public policy positions concerning science and technology if they take into account how rules and regulations may worsen the problem. They will be properly skeptical of claims made by those who supposedly speak in the name of "health and safety."

We all want to protect health and increase safety. We want to conserve the environment and help other people. But citizens must recognize that passing a law to solve one problem may create other challenges. The law may actually worsen the situation it tries to fix. More important, the law may blatantly disregard the values underpinning our society: individual freedom, property rights, and rule of law. People who understand trade-offs, who are able to appreciate the balancing of risks, will be well positioned to ensure the survival of our free society.

REFERENCES

Adler, J. (1995). *Environmentalism at the Crossroads: Green Activism in America*, Capitol Research Center, Washington, D.C., p. 17.

Advocates for Highway and Auto Safety (February 11, 1993) Comments to NHTSA. Docket No. 74-14-N79-006.

American Automobile Manufacturers Association (February 12, 1993). Comments to NHTSA. Docket No. 74-14-N79-023, p. 2.

Anderson, C. (1991) Cholera epidemic traced to risk miscalculation. *Nature*, 354:255.

Babbitt, B. (1994). The triumph of the blind Texas salamander and other tales from the endangered species act. *E Magazine* 5, No. 2 (April 1994), 54–55.

Bean, M. (1994). Speech from the U.S. Fish and Wildlife Service's Office of Training and Education Seminar Series, Marymount University, Arlington, VA, November 3, 1994. (This seminar was closed to the public; transcripts were obtained only by filing a request under the Freedom of Information Act.)

The Biggest Single Step to Curbing Global Warming (Undated). Statement by Sierra Club, Washington D.C.

Breyer, S. (1993). *Breaking the Vicious Circle: Toward Effective Risk Regulation*, Harvard University Press, Cambridge, MA, pp. 16, 23.

Claybrook, J. (1997). Air bags save lives; I still back them. Letter to the editor, *Wall Street Journal*, January 2, 1997.

Clinton, W., and Gore, A. (1992). *Putting People First*, Times Books, New York, p. 90.

CNN news report (1983, November 18).

Coalition for Consumer Health and Safety (1993). Comments to NHTSA, Docket No. 74-14-N79-021, February 12, 1993.

Competitive Enterprise Institute poll (1997). Conducted by *the polling company*, Washington, D.C., March 28, 1997.

Competitive Enterprise Institute vs. NHTSA, 956 F.2nd 321 (D. C. Cir. 1992), p. 322–324.

Competitive Enterprise Institute (1997, April). *Delisted and Endangered Species Act.* Washington, D.C.

Competitive Enterprise Institute. (1997, April). *The ESA's Dismal Record: The Failure to Recover Endangered and Threatened Species.* Washington, D.C.

Crandall, R. W., and Graham, J. (1989). The effect of fuel economy standards on automobile safety. *Journal of Law and Economics* XXXII:110–111.

Crews, C. W. (1996). Ten thousand commandments: A policymaker's snapshot of the federal regulatory state. Competitive Enterprise Institute Monograph, September 1996.

December 1979 *Car and Driver* comment printed in letters section of *Car and Driver*, December 1997, p. 16.

DeFalco, J. C. (1997). CAFE's Smashing Success: The Deadly Effects of Auto Fuel Economy Standards, Current and Proposed. Competitive Enterprise Institute Monograph, June 1997.

42 Federal Register 34, 292.

General Accounting Office (1988) Endangered Species: Management Improvements Could Enhance Recovery Program. RCED 89–5, p. 18.

Graham, J. D., and Wiener, Baert, J. (eds.) (1995). *Risk vs. Risk: Tradeoffs in Protecting Health and the Environment*, Harvard University Press, Cambridge, MA, pp. 1, 2, 89, 93, 90.

Insurance Institute for Highway Safety (1997) Letter to NHTSA regarding Docket No. 74-14, Notice 107, August 18, 1997.

Insurance Institute for Highway Safety (1993). Comments to NHTSA, Docket No. 74-14-N79-013, February 12, 1993.

Kazman, S. (1983). Deflating the claims of air-bag studies. Opinion page, *Wall Street Journal*, July 21, 1983.

Kazman, S. (1995). Amicus curiae brief of the Competitive Enterprise Institute in support of respondents, *Bruce Babbitt, Secretary of the Interior, et al. v. Sweet Home Chapter of Communities for a Great Oregon, et al.* pp. 4–5, 8.

Kazman, S. (1996). Naderites' nadir. Opinion page, *Wall Street Journal*, December 3, 1996.

Kumar, D. D., and Berlin, D. F. (1996). A study of STS curriculum implementation in the United States, *Science Educator*, 5(1).

Lambert, T., and Smith, R. J. (1994). The endangered species act: Time for a change. Center for the Study of American Business, St. Louis, MO. Policy Study, No. 119, March 1994.

Lewis, T. A. (1987). Searching for truth in alligator county. *National Wildlife*. p. 14.

Lomasky, L. (1995). Autonomy and automobility. Competitive Enterprise Institute Monograph, June 1995, pp. 10–12.

Lomasky, L. (1997). Sudden impact: The collision between the air bag mandate and ethics. Competitive Enterprise Institute Monograph, March 1997, p. 5.

Mesnikoff, A., and Pedery, S. (1996). Curbing our oil addiction to protect our health and environment. Intellectual Capital web site: <www.intellectualcapital.com/issues/96/10205/icpro.html>, December 1996.

National Research Council (1992). *Automotive Fuel Economy: How Far Should We Go?* National Academy Press, Washington, D.C., p. 75.

NHTSA brochure (1977) *Passive Vehicle Occupant Restraints.*

NHTSA Summary Report (1997). Relationship of vehicle weight to fatality and injury risk in model year 1985–93 Passenger Cars and Light Trucks. U.S. Department of Transportation, April 1997, p. 3.

O'Neill, B. (1997). Narrow account of air bag history. Letter. *Regulation* Spring: 20(2):4.

Seasholes, B. (1995). Species protection and the free market: Mutually compatible *Endangered Species Update* 12(4, 5):8.

Senate Commerce Committee Transcripts (1997). Hearing on air bags. January 9, 1997.

Struzik, E. (1992). Trouble back at the game ranch *International Wildlife* 24.

Sugg, I. C. (1993–1994). Caught in the act: Evaluating the Endangered Species Act, its effects on man, and prospects for reform. *Cumberland Law Review* 24(1):1–78.

Sugg, I. C. (1994). Rats, lies and the GAO: A critique of the General Accounting Office report on the role of the Endangered Species Act in the California fires of 1993. Competitive Enterprise Institute Monograph, August 1994, p. 4.

Tennessee Valley Authority vs. Hill, 437 U.S. 153 (1978).

Vivoli, M. (1992). Shoot, shovel, and shut up. *The Washington Times.*

Will Minivans Become an Endangered Species? National Center for Policy Analysis (1997). Brief Analysis No. 232, June 4, 1997.

CHAPTER 5

Thoughts about the
Evaluation of STS
More Questions than Answers

James W. Altschuld and David Devraj Kumar

INTRODUCTION

This chapter focuses on issues affecting the design and implementation of evaluation of STS (Science, Technology, and Society) programs, particularly accountability and national evaluation. Questions are raised about STS, and how its programmatic evaluation could be conceptualized and conducted. Also examined are important issues that must be considered in evaluation and a discussion of suggestions for implementing evaluation. The approach taken incorporates the critical, negative mask of the evaluator as described by Patton (1990), as well as the skeptical view suggested by Scriven (1973), with the intent not to dampen, but to bring to the surface the subtle issues that affect STS and the evaluation of STS programs.

In a social studies classroom, but not one in science (a critical point that will come up later), students were asked to identify major problems

James W. Altschuld, College of Education, The Ohio State University, Columbus, OH 42310
David Devraj Kumar, College of Education, Florida Atlantic University, Davie, FL 33314

Science, Technology, and Society: A Sourcebook on Research and Practice, edited by Kumar and Chubin, Kluwer Academic / Plenum Publishers, New York, 2000.

affecting society and then to delve into literature about a single problem. A student and one co-author chose to look at drug addiction. They asked: Where did it come from? What were its effects on individuals and society? How many people did it affect in society at that time? How did people kick the habit? How much did it cost society? What actions should society consider in regard to the problem area? Does taking drugs automatically lead to addiction?

Students were required to produce a paper and lead the class in a discussion of a variety of parameters around their chosen topic. It was not an easy assignment and it was not based upon the guidance currently available in the STS literature. (See the guidelines included in the NCSS position statement and guidelines in the *Social Education*, Volume 54, 1990.) The learning from this STS type experience, as retrospectively recalled and reconstructed over a very long period, was heavily slanted toward the societal part of the equation. The assignment might have had a much different slant if it were part of a biology or chemistry classroom where more of the science aspects would have been emphasized.

Yet, because the topic was one generated by students' own views and concerns about society, it was interesting and thought provoking. Indeed, a lot of learning occurred—if one can judge from a former student remembering an assignment several years later. To be specific, this course was similar to a curricular thrust in the Eight Year Study (1933–1941), whose main focus was the reform of secondary education in the United States. One emphasis in that study was a civics course that dealt with "Problems of Democracy." Problems of Democracy was intended to bring together students, representing different academic abilities and programs, to look jointly at the problems of society that affected everyone. So, although STS brings science and technology more clearly and appropriately into the contemporary picture of a global information age, it does have similar characteristics to what educators proposed long ago.

IMPORTANT ISSUES IN STS THAT AFFECT EVALUATION

While there are many potential issues in STS, the ones that will be discussed here relate to: the purposes of STS, place of STS within the curriculum, implementation of STS, teacher education, and extant evaluation procedures. See Table 1 for a brief overview of the issues.

Purposes of STS

From the Project Synthesis (Harms, 1977), Yager (1990) derived four main purposes of STS. In somewhat paraphrased form they are to:

Table 1. Some STS Issues Affecting Evaluation

Issue	Commentary
Purposes of STS	Many different purposes can be inferred for STS which, in turn could lead to many different evaluation emphases and activities.
Nature of STS	STS really combines three somewhat unique areas, and is interdisciplinary. Depending on its placement in the curriculum, the emphasis in STS instruction would be different resulting in varied outcomes. Interdisciplinary programs have strengths as well as weaknesses and are complex to evaluate.
Implementation of STS	How should STS be represented in the curriculum and implemented into instruction? Should it be implemented as separate STS units, integrated into other material, made an extension of another part of content, or as other options? STS programs at the upper elementary and middle school levels might look quite different than those in grades 9–12.
Teacher Education	To what extent are teachers, especially secondary science teachers, trained in STS programs, and do they have the requisite background necessary to implement them?
Extant Evaluation Strategies	The focus of STS on the integration of science, technology, and society concepts as viewed through an issue-oriented perspective is and will continue to be difficult to assess. The picture is even more complicated when the possibilities of varying goals for STS, patterns of implementation, and similar other factors are taken into consideration.

- prepare students to use science for improving their own lives and as a corollary to be able to better understand and cope with an increasingly technological society;
- enable students as they progress through life to deal in a responsible manner with technology–society issues;
- identify a body of knowledge that would enable them to deal with science–technology–society issues; and
- acquire knowledge and understanding about career opportunities in the field.

Hurd (1985) and Yager (1990) argued that STS would change the traditional ways of teaching science. Classrooms would be driven by exploration of current science and technology related problems facing society. Students would recognize and attend to social, technical, political, and humanistic factors. Instruction would tend to be based on problems chosen by students and on the interests of students, who would be motivated by working with current, personal and real world STS issues. Science instruction would, of necessity, have to incorporate hands-on techniques into the learning environment. With the progress in information technology, opportunities for students to seek and critically analyze the vast array of data and information currently available would increase, helping to develop intellectual skills such as decisionmaking, problem solving, knowledge synthesis, and ethical judgment. The teacher's role would be altered to be that of an active participant in examining societal concerns from the multiple dimensions of the three components of STS (Science, Technology, and Society) and from a wide array of issue perspectives.

Obviously, as Yager (1990) pointed out, there is much of value in STS, but its purposes give rise to many questions. Societal issues by their very nature encompass beliefs, values, aspects of aesthetics, economics, and so forth (Ramsey, Hungerford, and Volk, 1990). When looking at STS from this perspective one could wonder about its relationship to social studies (and history) and technology education, and about the background and skills of teachers dealing with such diverse content. Teachers are expected to approach an issue from the vantage point of the field in which they were trained and socialized. Thus, it should be noted that content emphasis in STS education would, to some degree, be a reflection of teacher specialization.

Given that the purposes STS are broad and provide considerable latitude for interpretation and implementation, subtle shifts in meaning or purpose potentially will occur. For example, Heath (1990) emphasized the idea that the goal of STS would be for students to apply their science and technological skills to the making of decisions, both personal and public. Wraga and Hlebowitsch (1990), and Remy (1990) echo similar perceptions of STS.

In a slightly different vein, although Rubba (1990), a science educator, identified ways in which social studies and science teachers could collaborate, he also noted that STS allowed students to apply their learning (presumably in science) to real life problems. Following this line of thought, some science educators might see STS as a way to enhance the relevance of the science curriculum to make science more related to forces and events that shape students' lives.

Interestingly, Wiesenmayer (1988, as reported in Rubba) conducted a study of a middle school STS program for 7th grade life science courses. The STS classes achieved significant pre to post-test gains for the three dependent variables, and their results were also significantly better on two of the three variables as compared to those of students in traditional life science courses. However, the traditional classes did somewhat better on life science concepts.

This finding is striking for it reveals different ways of thinking about STS outcomes. Is achievement in STS relationships and interdependencies an indication of understanding of science and technology implications for society? Is it to enhance the quality of science learning? Is STS possibly displacing some of the time devoted to science instruction for a related but somewhat different subject matter? If so, would the possible gain in relevance offset the differential in science content learning for the traditional classes, particularly at the upper secondary levels with their even heavier emphasis on science content? Therefore, should there be greater stress on STS in grades K–8 (or 9)?

Another question here deals with how STS might be affected by an emerging problem in science education. Based on international science test results, the United States does not fare well when compared to a number of other countries (U. S. Department of Education, 1997). One postulated reason is that the science curriculum is a "mile wide and an inch deep." (Note: Other reasons, such as the fit of test content to varying curriculums, teaching to the test, and better science education programs elsewhere, could also account for the results. Or as others point out, perhaps the notion of comparative thinking may not make as much sense as obtaining agreement on a criterion or domain referenced set of standards of what is important for U. S. students to learn.)

A mile wide and an inch deep refers to the many topics taught at a surface level in the United States whereas other countries teach a noticeably smaller set of topics across the grades which are studied in much greater depth. If a dramatic shift in the organization and teaching of science education takes place, how will the STS reform factor into it? If STS and certain aspects of issue exploration are part of the development of a new curriculum, will it be possible to embed STS type test items into international tests so that fair comparisons can be made?

Nature of STS

It could be argued that even the terms in the title of STS are in an arbitrary order that is open to challenge. The concept could easily be conveyed by society, science, and technology or by the words in some other order. This may seem trivial but on close inspection it may not be.

Science by its very nature and stance demands a neutrality. Traditionally, scientists are taught to steer away from political and societal forces. Did the United States of America enter the scientific exploration of space in the late 1950s because of pressure and public clamor from the scientific community? Or perhaps was it more due to national political and security concerns about Sputnik? Does a long-term emphasis on heart research result from the pristine views of medical researchers or did it arise because President Johnson had heart problems? Even with all of the heartfelt concern for those who suffer from AIDS and the terrible sadness of the disease and with all of the clamor for more research on AIDS, heart disease is still the number one health problem in the country. This fact often gets minimized in the political battle for funding.

According to May (1992), to comprehend and teach about such issues requires sociological understandings, aesthetic appreciations, historical perspectives, and the like. Further, she observed that there will be problems associated with the placement of STS instruction including:

- the "swamp" or what May referred to as the misperceptions and misunderstandings of other fields
- "turf" in which she questioned why does *Science* head the STS triumvirate especially when "the STS agenda is more visibly sociopolitical in character." (May, 1992, p. 76)

It seems doubtful that any one educator will be able to attain all the skills necessary to teach STS, and to implement and lead STS programs.

This is not to imply that many science educators: are not socially concerned individuals; do not see the necessity and importance of placing an issue orientation into science classrooms; and would not be able to implement and do a good job with STS programs. Nor does this deny or denigrate the importance of stressing the role of science and technology in modern life and understanding how they relate to change, political and economic forces. Modern science is acutely aware of how it must fight to make its case for support clear and cogent in the public forum and at the top decisionmaking levels in the administrative and legislative branches of government.

A sizeable portion of STS content and instructional approach may well reside in the social studies domain. A logical position could be taken in this regard. For example, some time ago an article appeared in the *Scientific American* about the characteristics of the Kiwi. The article concluded with a tongue-in-cheek suggestion that the Kiwi, based upon an examination of its peculiar traits, was "a mammal not a bird." By analogy and for the sake of argument, might not the same type of suggestion be extended to STS? And further, if STS lies more in the domain of social studies than science or technology, would the outcomes and expected results be different? Considering the interdisciplinary nature of STS, whether STS instruction occurs via the careful coordination of complementary content and activities in different subject matter areas or by teams of teachers is fraught with issues.

An example of just how difficult this seemingly straightforward idea can be in practice was observed in Project Symbiosis. The rationale of the project (Rossetti, 1992) was that the teaching of science within agricultural science classes would be enhanced by forming interdisciplinary teams of vocational agriculture teachers with teachers of science (physics, chemistry, biology, etc.). The project was conducted for two years with teams of teachers recruited from throughout the state of Ohio. The teams received training in teaming activities as well in regard to science materials that could easily be adapted for high school classrooms. The project was thoroughly evaluated during the period by means of survey data collected prior to the start of the project, during its implementation, and after its completion (Altschuld, 1993). Summaries prepared by the teachers describing their joint instructional ventures were also collected. In addition, each school site was visited at the end of the project and indepth interviews were conducted with participants.

An intent of this project was to place teaming at a natural nexus between two areas of the high school curriculum that appeared to be made for each other. To work in modern agriculture entails the use of many aspects of a highly advanced and rapidly changing scientific base (and for that matter a constantly evolving technological base). Unfortunately numerous problems were encountered that made teaming much harder to put into operation than project staff had anticipated.

Observations from the evaluation of the project as well as from its facilitation indicated that scheduling, a seemingly mundane consideration, was a major obstacle. Teachers are on fixed schedules, which are set long before the beginning of the school year. Thus it was mandatory that the project staff begin working with schools and administrators as early as the spring of the prior academic year for a project starting in the fall. Scheduling problems were encountered on two separate levels: teachers' planning periods and class periods for students who might benefit from new materials, methods, and interdisciplinary activities.

Obviously, it is advantageous to have common planning periods for involved teachers and additional released time. In Project Symbiosis it became immediately apparent that interdisciplinary work would require a serious and regular commitment of teacher time (jointly and individually) if it were to be successful. To go beyond anything other than surface types of activities (short superficial units that are not meaningful) necessitated that teachers: examine each other's syllabus; visit each other's classroom to get a first-hand sense of another person's teaching style; discuss their approaches to instruction; reflect on the types of students they are teaching; consider what content would best lend itself to working together, and so forth. Teachers ordinarily do not have opportunities to do so, particularly at the high school level. Many school environments are simply not set up to support team endeavors.

It is logical that the lessons learned about scheduling and the environment from Project Symbiosis apply to STS. Without a supportive context and strong administrative backing for the teaming effort, interdisciplinary programs will fall considerably short of achieving their outcomes. Since teachers have numerous preparations and serve different groups of students, thought must also be given to which students might benefit most from STS teaming. In some cases, the teachers in Project Symbiosis cited decisions related to combining classes and students as causing them difficulty.

Some other major problems in the project related to the physical proximity of teachers to each other (the closer in the building their classrooms were the better), prior experience with teaming (it turned out to be limited despite the average experience level of Symbiosis teachers, which was more than 12 years), and, surprisingly, by the lack of interest expressed by teachers (on questionnaires) to those parts of training dealing specifically with teaming as compared to those that focused on science concepts and hands-on activities that they could take directly back to their classrooms.

In regard to STS, interdisciplinary work could be exciting and has the potential for generating excitement in blending complex and diverse elements of instruction and content. In practice, as noted from Project Symbiosis, it may be hard to overcome the artifacts that reside in the details.

Implementation of STS

Overall. How are STS programs implemented? Kumar and Berlin (1996) in a nationwide telephone survey of state science supervisors reported the following implementation status of STS in K–12 education in the United States. Eight states required, nine recommended, and 20 states encouraged the use of STS in science eduction. Eight states reported some nebulous combination of required, recommended, and encouraged STS themes in their science curricula. While implementation of STS seems encouraging, there are still many unanswered questions.

In the Curriculum. In the wake of renewed national interest in science and technology one might expect an increased share of STS in science curricula. Kumar and Berlin (1998), in an analysis of 25 state science curriculum frameworks using 15 STS standards from the *National Science Education Standards* (National Research Council, 1996), found that 88 percent emphasized the standard science and technology in society. Three standards–environmental quality, science as a human endeavor, and nature of science and scientific knowledge—were stressed in nearly 50 percent of the state curriculum frameworks. The remaining 11 standards, such as history of science and historical perspectives, natural and human-induced hazards, were scarcely mentioned in the curriculum frameworks.

Although it is appealing to have STS mandated via state science curriculum frameworks, it may not be an effective curriculum strategy, because teachers are skilled professionals who have to tailor and adapt programs for their own classrooms and students. Rigid guidelines tend to fail. In regard to how STS content might enter the curriculum and classroom, Heath (1990) and Remy (1990) described three main mechanisms by which STS programs and content might be delivered: infusion into existing courses; extending existing aspects of a course or unit; and creation of a separate or specialized STS course.

An advantage of infusion is that learning in science, technology, and social studies courses can be enhanced by introducing and analyzing an STS oriented issue. Aspects of physics that focus on atomic energy can be examined from viewpoints such as: the demand for electric power; nuclear-generated power and its relationship to the current and projected overall energy requirements in a country such as the United States; the economics of nuclear energy; gaseous emissions; NIMBY (not in my back yard) concerns; and nuclear plant accidents such as Chernobyl or Three Mile Island. When these types of topics are injected into physics courses, they could immeasurably enhance the relevance of the study of atomic power and generate a great deal of enthusiasm.

Another option is to add an STS unit at the end of a course or after concluding a non-STS unit. Students might apply prior learning in the context of using information and knowledge to examine the complexity of decision-making and competing viewpoints. In the sense of the extended unit, the classroom becomes a microcosm of societal debates that surround STS issues.

The third choice would be to offer a separate STS course so that issues could be studied in depth. Remy noted several disadvantages of this approach, including the demands to develop and teach content that is drawn from three separate areas and the displacement of student time from other classes. A note of caution: New courses might not be well received in what are, at times, already extensive and crowded lists of course offerings in American schools. Remy also discussed the idea of interdisciplinary instruction, cautioning against its use.

Clearly there are many ways of implementing STS instruction and combinations of curriculum strategies. Taking infusion as an example, which could be done in a superficial or an indepth manner, how would these variations within approach affect learning and what types of learning would be affected? When the permutations of delivery style are taken into consideration, the evaluation of STS becomes even more complicated. Results could not be easily interpreted without a solid understanding of program implementation. Without a description of the subtleties of implementation, results lose much in the way of meaning.

Levels. STS could be implemented at different levels within the educational system. It is unlikely that it would start earlier than 4th grade since it calls for integrating concepts across its three components. Students probably should be beyond the concrete level of development before getting into STS learning. However, Kumar and Berlin (1996) reported that STS education is implemented in 29 states (eg., Arkansas, Florida, North Carolina, Ohio) in grades K–12, in five states (eg., Iowa, Louisiana) in grades 6–12, and in six states (eg., Nebraska, Utah) in grades 9–12 or 7–12. According to Kumar and Fritzer (1998), in Florida, 16 percent of STS implementation is at the elementary school level, 47 percent is at the middle school level, and 35 percent is at the secondary school level.

Elementary grade teachers tend to have greater experience with the integration of concepts given that they routinely teach more than one subject matter. Integration may be more related to what they do everyday than for secondary level teachers who are specialized and departmentalized. In some aspects of STS programs, no matter how well-intentioned the teacher in science, technology, or social studies is, the complexity of topics might require the input of teachers from other disciplines. Moreover, when levels are taken into account, the question arises: What is the articulation of STS across the grades, and what should it be?

This question becomes paramount when looked at from the perspective of less is more and a science curriculum that could evolve into a mile deep and perhaps a few inches wide. Would evaluations change with an articulated STS program that focused on fewer topics and that was coordinated across levels? Would 11th and 12th grade teachers be aware of what was taught in STS at lower grade levels, how it was tied into curriculums and instruction at those levels, and how it would relate to what they would be doing?

Another problem is how teachers themselves feel about what is the appropriate placement for STS programs. Rhoton (1990) conducted a study of 7th–12th grade teachers in the state of Tennessee. Not surprisingly, very few of them felt that STS should be taught at the elementary school level, with the bulk of the responses loading on the middle and secondary grades. Some of the other findings of this study were that, in the current STS curriculum, students were neither independently investigating STS issues, evaluating alternatives to the problems inherent in the issues, nor acting on alternatives. In addition, seven questions on the survey dealt with difficulties the teachers perceived in integrating STS into the curriculum. All seven areas were rated as being major problems affecting implementation. Clearly the environment in which STS programs exist would be an important feature affecting their implementation.

Teacher Education

Inservice and preservice teacher education should equip teachers to handle the diverse body of content and the instructional needs of STS. The demands would vary according to whether a science, social studies, or technology teacher is implementing an STS program singly or in a group. How much preservice training is required? And how much inservice training would be required? In a national survey of state science supervisors in the United States, Kumar and Berlin (1996) found that 34 states have given inservice workshops addressing STS education. In a follow-up study of STS implementation in Florida, Kumar and Fritzer (1998) found that 31 percent of school districts provided training in STS for teachers. While inservice training in STS looks promising, overall information about preservice training in STS is lacking. The following are a few critical issues that teacher educators should bear in mind while addressing STS education.

Keeping the focus on science teachers (but recognizing that many of these same questions in modified form would apply to social studies or technology teachers), have they been socialized in their preservice training to teach in such an issue-oriented fashion? (Ramsey, Hungerford, and Volk, 1990, described an "Issue Analysis Technique" that would seem to be a critical and central element in STS programs.) Are science teachers trained to systematically teach students how to analyze what are often very complex issues? Are these teachers familiar with STS issues and do they understand that many issues are framed by concerns outside of science? How, for example, would they tactfully address problems in forestry in states such as Washington and Oregon where conservationists are warning about environmental dangers and resource problems at the same time as jobs and livelihoods are on the line?

Are preservice teachers trained to handle the values that could arise in regard to cloning, evolution (vs. creation arguments), the role that science played in the Holocaust, the use of gene technology to alter (and improve) plants and animals, the economic and political forces that are leading to reduction in the size of the South American rain forests, the dependence of the United States on fossil fuels for transportation, and the disproportionate use of energy in developed vs. underdeveloped nations? Are they prepared to help students analyze the nature of misperceptions and pseudoscience in relation to how they influence decisions made on an everyday basis and those made at legislative and higher levels affecting virtually all aspects of society? Do they reflect on their own perceptions of issues and why they hold them? Are they aware of ethical considerations in the use of scientific principles? Are they aware of career opportunities in STS, and are they trained to help students explore such careers? (One could even question whether teachers are aware of careers in science, let alone STS, with its complex union of three distinct areas.)

Although a handful of teacher preparation programs (e.g., the non-state-mandated Iowa Chautauqua Project at the University of Iowa) have stressed STS, there is very little generalizable evaluation data available to understand to what extent preservice science education programs stress STS nationwide. How would STS preservice training fit with the press of science teachers at the upper secondary level to gain more fundamental training in their specialized area of science? What supportive factors should be available to enhance the delivery of STS programs? Clearly, teacher training is an essential part of the successful delivery of STS. Considerations about teacher training and background along with knowledge of supportive, contextual factors would play a large role in STS programs, and, in turn, have dramatic effects on large scale evaluations of STS programs.

How is the school or school system encouraging and supporting inservice STS training/workshops? Are adequate resources allocated for teacher training? For example, Kumar and Fritzer (1998) found in a survey of STS in Florida, only a minute percentage of the districts responding provided any funding for STS materials. A serious concern here is that "although 31 percent of the 49 districts responding had provided teacher and/or supervisor training in STS, only 4 percent provided special funding, and only 8 percent made training available for supervisors. The obvious question this brings to mind is how so many were trained with so little availability of support" (Kumar and Fritzer, 1998, p. 16).

Extant Evaluation Strategies

Cheek (1992) developed a thoughtful overview of the state of STS evaluation. He noted that, as of the early 1990s, most standardized testing approaches did not include items designed to assess STS curricula and learning. As a set of general evaluation considerations, Cheek proposed that districts or schools evaluate their STS efforts by means of a rich array of strategies for determining student outcomes. The strategies might include: written reports, presentations, interviews, self evaluations, and so forth.

Cheek correctly pointed out that STS calls for students to integrate content and ideas and to view an issue from many perspectives, even those in opposition to each other. Such integration is the very essence of the concept of STS and is similar to how evaluation is approached at the graduate level via qualifying or general examinations.

The implication is not that STS programs in K–12 education are the equivalent of those at the graduate school level. Instead, the argument here is that a consistent process of integration is at the heart of the matter. By analogy, on a general examination in education a student might be asked to deal with the pros and cons of an issue, the rationale behind it, and the nature of why some power groups are fighting for the topic whereas others oppose it. The individual might even be asked to articulate a personal stance on the issue and to defend it in light of current understandings. Examples of issues could be charter schools, statewide funding and control of schools, national certification of teachers, national educational tests, the STS movement, and other current and contentious concerns facing the educational establishment.

Certainly, students are expected to have command of a wealth of facts and resources about an issue and to be able to demonstrate that their discussion is based upon a serious exploration of it. While such knowledge is necessary, is it sufficient for integration? If it were, then simply test or look for evidence of knowledge. In STS, the learner is required to go beyond factual knowledge to understand the values and beliefs that drive decisions related to scientific and technological development. The learner must be able to comprehend the advantages and disadvantages (economic and otherwise) of conflicting positions, and to pull together pro and con arguments that are derived from multiple sets of highly varied information. These are not easy tasks.

The suggestions offered by Cheek (1992) are helpful in thinking about the integration required by STS. At the same time that he described examples of multiple choice items designed to test STS content, Cheek also suggested that there was dissatisfaction with the limitations of such items for determining the extent of STS learning. Other ways to approach measuring STS include open-ended assessment items, essay examinations, performance-based assessments, and the use of portfolios. Cheek concluded his review of evaluation by briefly mentioning new instruments under development at the time of his article and by urging educators to consider an "integrated STS assessment" derived from a range of assessment techniques.

In principle, there are a number of problems beyond those explained by Cheek. Multiple measures are desirable in a generic sense but there are difficulties in implementing them and, in turn, in analyzing and interpreting results. This would be especially true when evaluating an overall STS program in a school district. Analogously, Altschuld and Witkin (2000) have looked at the use of multiple measures in assessing educational and social needs. In many instances, it was apparent that the results simply did not agree, and in one case, they actually disagreed. These authors also observed that costs and the skills of the evaluators who must be proficient in a series of methods posed another problem.

A final major problem would be describing what is evaluated. STS programs will vary in duration, strength, emphasis, style of implementation, content, and in many other dimensions. So even if tests were employed as the main evaluation technique, one would have difficulties in saying what STS was and what in it potentially led or contributed to the observed outcomes.

A fundamental premise of program evaluation is that the evaluators are able to describe the entity being evaluated. It is the *sine qua non* condition of evaluation. If science education in the United States has been criticized for being a mile wide and an inch deep (a finding partially based on an indepth description of the curriculum), then the STS component of science education must also be depicted in detail.

SUGGESTIONS FOR EVALUATION

Obviously, the evaluation of STS will neither be an easy nor simple undertaking. When looked at collectively the issues raised in previous sections might lead to questions of how STS should be evaluated. Some educational efforts and programs are inherently difficult to evaluate (e.g., sending children to summer camp), although it is clear they have value. Conversely, STS could drastically alter instruction and content wherever it is taught. It could (and indeed it probably would) displace some content due to the insertion of STS content into a course. One argument in support of doing this is that STS will make the science, technology, and social studies content more meaningful and relevant. Yet, others would argue (as put to one of the coauthors by his wife, a prominent heart cell researcher) that the call for relevance is good, but it comes at the sacrifice of the precious little time and opportunity needed for the study of science. What are the gains to be realized and what are possible losses that might result from moving toward an STS stance? If the funds and resources expended for STS become sizeable, the concern regarding evaluation and overall accountability will become more pronounced.

In Table 2, four suggestions for STS evaluation are shown. They are: description of STS implementation; evaluation of the context; a general

Table 2. Four Approaches to Evaluating STS Programs

Evaluation Approach	General Comments
Description of STS Programs	Rich description of STS programs is needed. It is suspected that programs would fall into classes or groups that would enable evaluators to better interpret STS results.
Evaluation of Context	The context could be a major determiner of the success or failure of STS programs. The context, particularly as it relates to the acceptance of a change such as STS, is important.
General Meta Analysis	If enough studies had been completed and data for calculating effect sizes were available, it would be possible to obtain an overall sense of the impact of STS programs. The analysis would most likely be based on standardized tests or other similar measures.
Specified Meta Analysis	If the above meta analyses were computed from studies that also contained descriptors about program types, then it would be informative to look at results and outcomes by types of programs or other descriptors of programs.

meta analysis; and a specified meta analysis. They are explained in some detail below.

Description of STS Programs

A first step for evaluation would be to conduct a descriptive analysis of a sampling of STS programs throughout the country. The goal would be to derive a typology of how STS is being implemented (subject matter focus, length of treatment, style of implementation, team teaching and interdisciplinary focus, displacement within the curriculum, etc.). In other words, what does STS mean when it is not framed from a "theoretical" perspective, but instead operationalized in real world settings?

This is not a new idea. In 1970, Cohen examined federal programs and suggested that because so many variations could comprise a single program, then a description of program implementation and type would be an imperative aspect of evaluation. Warmbrod (1977), and Altschuld and Downhower (1980), though working in different contexts, came to the same conclusion. The evaluation of federal programs for drug-free communities also ran into the need to describe program types. A suspicion is that, in STS, natural groupings of how programs are implemented would be evident. Having descriptions of these groupings would be extremely informative for understanding what STS is in practice and how types of implementation strategies might eventually relate to outcomes achieved.

Evaluation of Context

In 1995, Altschuld and Kumar proposed a model for the evaluation of science education programs. This model's main feature was its emphasis on the evaluation of the context surrounding a program from the beginning of conceptualization, through development of the program, to its full-scale, long term installation in a school system. In an application of this model to evaluate an NSF-funded, innovative technology-based science teacher education project, Kumar and Altschuld (1997) found that the degree to which the context supports innovation and change, in general, and a specific innovation, in particular, were critical factors leading to success or failure of change. Smith and others (1993) and Goodway-Shiebler (1994) have come to similar conclusions regarding change in health education programs.

This kind of thinking would seem to apply equally well to STS. If, for example, a group of teachers wanted to implement an interdisciplinary STS program, the nature of the environment would play a key role. Administrators could help to arrange schedules so that adequate planning could occur, or they could provide time and funding for teacher training. Community resources might have to be located and solicited to help in program delivery. Local university faculty might be asked to provide inservice training for STS and to assist with the facilitation and evaluation of the endeavor.

Context information is frequently obtained through time-consuming and costly qualitative procedures. Despite the time and cost, the collection of context information during the process of describing STS programs is strongly encouraged.

General Meta Analysis

In essence, a meta analysis seeks to determine the average effect size across studies of a type of treatment or program. If sufficient studies were available it might be worthwhile to conduct a meta analysis of STS programs. Is there an overall effect, what is its size, in what direction is it, and how stable is the result (i.e., how many studies led to the meta analytic finding)?

Meta analyses are not without their own difficulties. In the case of STS, most of the data would be from studies using well established, standardized tests or promising STS testing procedures that have been developed in the 1990s such as some of those described by Cheek. The richness of the integration of concepts that is so central and important to STS may not be measured as well by tests and multiple choice items as it is by other techniques previously referred mentioned. Undoubtedly, qualitative outcome data would be useful and should be collected. Even so, meta analyses provide science educators with evidence and an indication of the impact of STS programming.

In conducting meta analyses or other types of studies, it would be wise for evaluators to look not just for STS-specific outcomes, but also for different outcomes. STS could (not necessarily) come at the expense of other learning. STS results may have been achieved, but at the same time how did science or technology outcomes compare to non-STS classes?

One more consideration is that meta analyses, and probably most other studies of STS, are fairly current or present-time oriented. Current learning is critical and serves as the basis for subsequent, years-later action and decisionmaking. (While it would be problematic to collect information about how those individuals trained in STS courses make STS decisions five, ten, fifteen years into the future, a long term study of the effectiveness and impact of STS instruction will eventually be needed.) Will they be more reflective as decisionmakers about STS issues? Will STS learners try to understand alternative value positions and beliefs as they relate to how others view the issues? Will they be able to identify misperceptions that they perhaps might hold?

Specified Meta Analysis

Ideally, it seems that if STS program descriptors and context data had been collected, then the meta analysis would be greatly enhanced. These variables could be entered into the analysis so that it would help us to understand effect size in terms of environmental support and program type. This information would permit STS proponents to make recommendations about how to implement STS programs. What works? What doesn't? How much support does it take? Do team efforts work better than those of individual teachers? Clearly, description of STS programs, evaluation of the context, and general meta analysis all contribute to this more probing, specified meta analysis.

SUMMARY

If the issues and concerns raised in this chapter foster debate about STS and its evaluation, then the chapter has been successful. There are many good ideas in STS. Certainly as a reform it has engenderd much enthusiasm in a number of fields. However, this enthusiasm must be tempered by the questioning attitude that STS itself strives to produce.

Approaches to evaluating STS programs may include gaining a rich description of the program and its context, and general as well as specified meta-analyses. Contextual factors include an environment conducive to innovation and development in STS instruction, strong administrative interest and commitment to further the cause of STS; and faculty interest and commitment to be actively involved in STS activities and to integrate STS themes into their own teaching practices.

The purposes of STS education must be clearly identified to devise focused evaluation strategies. Clearly delineated information about the nature of STS programs and their implementation by grade levels is critical to fully understanding and evaluating the effectiveness of STS education. The issue-oriented interdisciplinary nature of STS makes designing structured evaluation strategies a challenge.

Teacher education is similarly critical to the successful implementation as well as evaluation of STS programs. As Brinckerhoff (1985) pointed out, teachers often show reluctance to use STS instruction due to "(a) a lack of class time in an already overcrowded syllabus, (b) inadequate knowledge of the facts surrounding debatable issues, and a lack of usable materials in print . . ." (p. 4). In a review of STS literature Kumar and Berlin (1993) suggested that "studies to investigate the role of STS themes and issues in preservice science teacher education programs should be conducted to help educators redesign preservice science teacher education programs and prepare STS literate teachers" (p. 79).

Whether STS is a valid approach to teaching and learning science, technology, and society depends upon what STS has accomplished so far in education. Over two decades of debate has brought some visibility to it as a curriculum reform. STS themes are slowly permeating state and school district science and social studies curriculum frameworks. However, hard core evaluation data is needed to argue that STS will continue to be an important part of education in the next millennium. The STS community of scholars should consider focusing a portion of their efforts on designing and conducting valid evaluation strategies that would shed light on where STS is now, where it should be in the future, and how policies are to be developed to guide the course of STS education in the next century.

REFERENCES

Altschuld, J. W., and Witkin, B. R. (2000). *From Needs Assessment to Action: Transforming Needs into Solution Strategies.* SAGE Publications, Thousand Oaks, California.

Altschuld, J. W., and Kumar, D. D. (1995). Program evaluation in science education: The model perspective. In O'Sullivan, R. G. (ed.), Emerging Roles of Evaluation in Science Education Reform, *New Directions for Program Evaluation, 65,* 5–17.

Altschuld, J. W. (1993). Evaluation of project symbiosis: An interdisciplinary science education project. *Cognosos* 2(3):4–5.

Altschuld, J. W., and Downhower, S. G. (1980). Issues in evaluating the implementation of public law 94–142. *Educational Evaluation and Policy Analysis* 2(4):31–38.

Brinkerhoff, R. F. (1985). New techniques for teaching societal issues in school science course: Abridged versions. *SSTS Reporter* 1:4.

Cheek, D. W. (1992). Evaluating learning in STS education. *Theory into Practice XXXI* (1):64–72.

Cohen, D. K. (1970). Politics and research: Evaluation of social action programs. *Review of Educational Research* 40(2):213–238.

Goodway-Shiebler, J. D. (1994). The effect of a motor skill intervention on the fundamental motor skills and sustained activity of African-American preschoolers who are at-risk. *Dissertation Abstracts International* 55:3781.

Harms, N. C. (1977). Project Synthesis: An interpretive consolidation of research identifying needs in natural science education. A proposal prepared for the National Science Foundation, University of Colorado, Boulder, CO p. 434.

Heath, P. A. (1990). Integrating science and technology instruction into the social studies: Basic elements. *Social Education* 54:207–209.

Hurd, P. D. (1985). A rationale for a science, technology, and society theme in science education. In Bybee, R. W. (ed.), *Science Technology Society. 1985 Yearbook of the National Science Teachers Association*, National Science Teachers Association, Washington, D.C., pp. 94–101.

Kumar, D. D., and Berlin, D. F. (1998). A study of STS themes in state science curriculum frameworks in the United States. *Journal of Science Education and Technology* 7(2):191–197.

Kumar, D. D., and Altschuld, J. W. (April). Contextual variables in technology-based science teacher education. A paper presented at the International Conference of the Society for Information Technology and Teacher Education, Association for the Advancement of Computing in Education, Orlando, FL, April, 1997.

Kumar, D. D., and Berlin, D. F. (1996). A study of STS curriculum implementation in the United States. *Science Educator* 5(1):12–19.

Kumar, D. D., and Berlin, D. F. (1993). Science-technology-society policy implementation in the USA: A literature review. *The Review of Education* 15:73–83.

Kumar, D. D., and Fritzer, P. J. (1998). A study of science-technology-society education implementation in the state of Florida. *Journal of Social Studies Research* 22(1):16–20.

May, W. (1992). What are the subjects of STS-really? *Theory into Practice XXXI* (1):73–83.

National Research Council (1996). *National Science Education standards*, National Academy Press, Washington, D.C. p. 262.

Patton, M. Q. (1990). Workshop presentation. Aannual spring workshop of the Ohio Program Evaluators Group, Columbus, OH, May 1990.

Ramsey, J. M., Hungerford, H. R., and Volk, T. L. (1990). Analyzing the issues of STS. *The Science Teacher* 57:61–63.

Remy, R. C. (1990). The need for science/technology/society in the social studies. *Social Education* 54:203–207.

Rhoton, J. (1990). An investigation of science-technology-society perceptions of secondary science teachers in Tennessee. *School Science and Mathematics* 90(5):383–395.

Rossetti, R. (1992). Improving teachers' ability to teach science. A proposal to the W. K. Kellogg Foundation. The Ohio State University, Columbus, OH. p. 23.

Rubba, P. A. (1990). STS education in action: What researchers say to teachers. *Social Education* 54:201–203.

Scriven, M. (1973). The methodology of evaluation. In Worthen, B. R. and Sanders, J. R. (eds.), *Educational Evaluation: Theory and Practice*. Wodsworth Publishing Company Belmont, CA, 60–104.

Smith, D. W., Steckler, A. B., McCormick, L. K., and McLeroy, K. R. (1993). Teacher's use of health curricula: Implementation of growing healthy. Project SMART and the teenage health teaching modules. *Journal of School Health* 63(8):349–354.

U. S. Department of Education (1997). Attaining excellence: TIMSS as a starting point to examine U. S. Education (introduction to TIMSS). Pittsburgh, PA: U. S. Government, Superintendent of Documents (p. 74).

Warmbrod, J. R. (1977). Evaluation research in vocational education. *Beacon: The Newsletter of the American Vocational Evaluation Association* 6(1):1–11.

Wiesenmayer, R. (1988). The effects of three levels of sts instruction and traditional life science instruction on the overt citizenship behavior of seventh grade students. (Doctoral dissertation, The Pennsylvania State University), p. 312.

Wraga, W. G., Hlebowitsh, P. S. (1990). Science, technology, and the social studies. *Social Education* 54:194–195.

Yager, R. E. (1990). STS: Thinking over the years. *The Science Teacher* 57:52–55.

CHAPTER 6

Science, Technology, Society, and the Environment
Scientific Literacy for the Future

Kathleen B. deBettencourt

INTRODUCTION

Environmental education and the science–technology–society movement came into being at about the same time and in response to a similar concern: Education must do more to develop an informed citizenry capable of making decisions about current problems, particularly issues involving science and technology.

In 1970, several universities, including Cornell, Pennsylvania State, and Stanford, began programs on what is now referred to as science–technology–society (STS) (Yager, 1993). In 1977, the Project Synthesis, National Science Foundation-funded science curriculum research effort, made STS one of its five focuses. They identified several goals, including preparing students to "use science for improving their own lives and for coping in an increasingly technological world" and to "deal responsibly with

Kathleen B. deBettencourt, Environmental Literacy Council George Marshall Institute, Washington, D.C. 20006.

Science, Technology, and Society: A Sourcebook on Research and Practice, edited by Kumar and Chubin, Kluwer Academic / Plenum Publishers, New York, 2000.

technology–society issues" (Harms, 1977 cited in Yager and Roy, 1993, p. 8). Following the work of Project Synthesis, the National Science Teachers Association (NSTA) began its Search for Excellence, identifying STS as one of the search areas. In 1984, NSTA unanimously adopted a statement recommending that all students in American high schools receive exposure to STS topics.

The environmental education movement also had its beginnings in the early 1970s. It grew, as one of the founders of the field reports, "out of growing discontent with how we (esp[ecially] Americans) were treating the air, water, plants, soil, and animals, and how schools were preparing future citizens to make intelligent decisions about the environment" (Knapp, 1996, p. 12). In 1969, William Stapp provided a definition of environmental education that is the basis for many subsequent statements of the purpose of the field:

> Environmental education is aimed at producing a citizenry that is knowledge-able concerning the biophysical environment and its associated problems, aware of how to help solve those problems, and motivated to work toward their solution. (Stapp, 1969, pp. 30–31)

This definition is echoed in the NSTA's 1990 position statement that "Basic to STS efforts is the production of an informed citizenry capable of making crucial decisions about current problems and issues and taking personal actions as a result of these decisions" (cited in Yager, 1993, p. 3).

STS and environmental education share more than purpose; they share subject matter. Peruse any middle school or high school science textbook and you will see "science–technology–society" sidebars on global warming, pollution, and other environmental topics. Of the eight specific areas of concern that Project Synthesis identified as characterizing STS, five relate to the environment: energy, population, environmental quality, use of natural resources, and effects of technological development.[1] Indeed, Chiang-Soong's study of American textbooks indicates that issues concerning environmental quality and natural resources predominate as STS topics (Chiang-Soong, 1993). Because of STS' many linkages, some environmental educators argue that it *is* environmental education (Rubba and Wiesenmayer, 1998; Disinger, 1986; Volk, 1984).

Environmental issues predominate as STS topics for the same reason that schools are flooded with environmental education resources. Environmental issues are pervasive in the media and are increasingly the subject of

---

[1] The other areas are human genetic engineering, national defense, space, as well as sociology of science. Many environmental educators would also include these topics as concerns of environmental education.

public concern, especially among students. If one of STS' goals is to help students understand the relevance of science to their everyday lives, then environmental issues are useful areas to explore. Environmental studies, moreover, hold great promise for drawing students to study science further. And, of course, decisions about the natural resources and environment will be on the public agenda for generations to come. It is critical that students be prepared to participate knowledgeably in these discussions.

There is evidence, however, that after 25 years of incorporating environmental issues in the curriculum, students understand very little about the environment. Surveys of environmental knowledge among students are sobering. A 1992 Roper poll, for example, tested teens' knowledge on issues such as air and water pollution, solid waste, and conservation (Roper, 1992). Roper had previously administered a similar quiz to adults; the report noted, "Our expectations, frankly, were that teens would do better [than the adults] in this exercise, but these expectations proved to be optimistic" (p. i). Teens, in fact, scored slightly lower than adults on the quiz, even on topics frequently covered in environmental education curricula, such as waste management, acid rain, pollution, and biodiversity.

A 1990–1991 study of over 3,000 students in 30 secondary schools in New York State identified a similar lack of knowledge. Remarkably, students who reported studying environmental science tended to score lower on the knowledge test than those who had not (Hausbeck, Milbrath, and Enright, 1992).

A national survey of high school students' environmental knowledge (based on responses to environment-related questions designed for National Assessment of Education Progress testing in science and mathematics) found that, although a majority of students could identify sources of environmental problems (e.g., fossil fuels, smokestacks emitting sulfur dioxide), relatively few could identify the consequences (Gambro and Switzky, 1996). For example, only 43.5 percent of the high school seniors in the sample could identify three consequences of the greenhouse effect. The study concluded "most high school seniors possess an elementary comprehension of environmental problems and lack the necessary understanding to go beyond the common recognition of an issue and use their knowledge to grasp the consequences of environmental problems or offer solutions for those problems" (Hausbeck, Milbrath, and Enright, 1992, p. 22).

Despite all our efforts, how well we are preparing students to understand environmental issues? To address that question, the Independent Commission on Environmental Education, a project of the George C. Marshall Institute, brought together a panel of scientists, economists, and educators to perform an extensive study of commonly used textbooks and environmental education resources.[2] Commission members came from diverse fields and viewpoints. Members included one of the founders of the field of environmental education, as well as eminent ecologists, economists, and other experts in environmental science. The commission's report, *Are We Building Environmental Literacy?*, was published in April 1997.

The commission found textbooks and teacher's guides available at all levels that provided a challenging and thorough introduction to environmental topics. Many of the texts reviewed, however, were flawed with factual inaccuracies and dated scientific theories, or gave only superficial explanations of complex environmental issues.

Certainly, factual errors in textbooks are not a new problem. The National Research Council's 1990 study of biology education in American schools, noted that, "this fundamental criterion [factual accuracy] is often violated" (National Research Council, 1990, p. 57). Coverage of environmental issues, however, present special challenges. Environmental science is a complex multidisciplinary study that draws from the natural and the social sciences, including biology, chemistry, economics, and political science. This is a field in which knowledge is growing rapidly, and in which there are competing theories and interests.

Yet these same aspects make environmental issues particularly interesting for students. The environment permits students to integrate geography, science, social science, history, and other courses in an investigation of issues of immediate relevance. Unfortunately, most texts miss this opportunity. The following is a summary of some of the topic areas reviewed by the commission and its findings.

TOPICS REVIEWED BY THE INDEPENDENT COMMISSION AND FINDINGS

A Hot Issue: Global Warming

Global warming is frequently featured as an STS issue and is a significant subject in environmental education materials. Climate change presents an excellent opportunity to combine science concepts, such as climate and weather, with a discussion of political and economic considerations. It is apparent, though, that authors of many textbooks do not themselves have a clear grasp of the science involved.

---

[2] The George C. Marshall Institute is a nonprofit research group interested in education, science, and technology and their impact on public policy, particularly environmental policy and national security. The institution receives its funding from private foundations. It receives no industry or corporate support.

One common problem in many textbooks stems from confusing the terms "global warming" and the "greenhouse effect." In part, this is understandable, because in the late 1980s when concerns were first raised, these terms were often used interchangeably. The greenhouse effect, though, is a well understood phenomenon, an important part of earth's systems. The greenhouse layer retains enough of the sun's heat to warm the atmosphere to a temperature sufficient to sustain life. Global warming, on the other hand, refers to the theory that increased greenhouse gases produced by human activity will *enhance* the natural greenhouse effect and raise global temperatures.[3]

Only some texts explain the greenhouse effect. Yet without this concept, students will have a hard time understanding global warming. Often texts fail to explain the greenhouse effect is the result of naturally-occurring gases in the atmosphere, including carbon dioxide, methane, nitrous oxide, ozone, and particularly, water vapor. Much of the uncertainty in predicting future climate relates to scientists' currently limited understanding of the role of water vapor and clouds in atmospheric dynamics. It is not helpful to discuss, as many texts do, only carbon dioxide. And an unfortunate number of textbooks and teacher's guides confuse key terms and mislead by oversimplification. Only a few global warming discussions in the texts reviewed mention that projections are based on computer models or even discuss the uncertainties inherent in predicting climate change, in light of so many variables that are poorly understood by professionals working in this field.

For example, Holt's *Science Plus*, explains the greenhouse effect in its chapter on climate, but then refers students to an STS sidebar which begins, "*No one knows whether there is a greenhouse effect* or whether there will be global warming" (emphasis added) (Morrison *et al.*, 1993, pp. 408–409). This statement will baffle those students who were paying attention to the previous chapter which explained the greenhouse effect's important role in earth's atmosphere.

Addison-Wesley's environmental science textbook (among others) refers to carbon dioxide as a "pollutant," although carbon dioxide generated from human sources is indistinguishable from the vast amount produced by natural sources (Bernstein, Winkler, and Zierdt-Warshaw, 1996, p. 366). As Kempton, Boster, and Hartley (1996) note in their study of environmental values in America, viewing greenhouse gas emissions as a pollution problem is misleading, because the solution is assumed to be pollution control technology, such as scrubbers, yet no such technology exists for "filtering" carbon dioxide.

---

[3] Indeed, because changes predicted by global climate models effect weather and other systems, scientists use the term global climate change.

Prentice Hall's *General Science* includes a vignette which describes (with illustration) a tour by glass boat of a submerged city through towering skyscrapers. After explaining this is probably an effect of global warming, the text asks:

> Can anything be done to reverse the earth's greenhouse effect? Not much, say
> scientists. We depend too much on fossil fuels for our energy needs. Then what
> can we do? Get ready is the answer. Make plans to hold back the sea. (Hurd
> *et al.*, 1992, p. 403)

This dramatic presentation will no doubt get students' attention, but what does it teach them about the role of science in society? The accuracy one must expect in a science textbook is lost in the attempt to stress a point. We do not want to "reverse the greenhouse effect"; without it earth would be a cold place. The presentation considerably overstates the rise in sea levels predicted by even the most pessimistic global climate models, and attribute to scientists an omniscience that mischaracterizes the state of current knowledge about climate impacts.

Textbooks often emphasize personal actions that students can take to conserve energy. LeBel's *Environmental Science: How the World Works and Your Place In It*, for example, lists measures that should be taken to reduce carbon emissions, ranging from using mass transportation, to recycling, to planting trees (Person, 1995). These suggestions are salutary and teach good citizenship. However, students will not understand the scope of the issue unless they also consider that considerable costs will be incurred in transforming a transportation system that relies on fossil fuels for approximately 90 percent of its energy, or that any efforts to reduce greenhouse gas emissions in the industrialized countries will be overtaken in the next few decades by greenhouse gas emissions arising from rapidly industrializing nations, especially in Southeast Asia. Particularly in upper level texts and materials, students ought to be introduced to economic considerations.

Global climate change is one of the most interesting and important science-based social issues of our time. Students will be called upon to consider public policies and, perhaps, to make personal sacrifices. More substantive and accurate coverage of this important topic should be developed for students, particularly those in upper level courses.

Acid Rain

Coverage of other topics is also flawed by incomplete or misleading presentations. Acid deposition, for example, is another example of an environmental issue that permits students to see how the chemistry they are learning in science classes has relevance to real world concerns.

However, in many texts the policy aspects are covered in more detail than the science.

These discussions are seldom linked to an explanation of the chemical process that results in acidic deposition. Units on acid rain are typically accompanied by an "experiment," although these activities are designed more to illustrate a point than for scientific inquiry. For example, *Science Insights*, a middle school general science text, concludes its brief presentation of acid precipitation with an experiment using two sets of seeds, one of which is watered with "normal" water and one which is watered with water mixed with vinegar (DiSpezio *et al.*, 1996, p. 619). The text does not instruct students in testing the pH of the normal water nor does it say whether the "normal" water should be tap water, distilled water, or collected rain water. The text also does not explain that rain water is naturally acidic, much less describe what a pH scale is and how it is derived.

The results of this missed opportunity are evidenced by American students' performance on the Third International Mathematics and Science Study (TIMSS). Only 32 percent of American students could correctly answer a multiple choice question that asked them to identify one of the principal causes of acid rain, despite the widespread coverage of this issue in texts for a number of years (TIMSS, 1997, chapter 3, table 3.5). Students' failure to comprehend the source of an environmental problem means they will be less capable of understanding the costs and benefits of proposed solutions.

Acid rain is also an area in which scientific theories have changed since the issue was widely reported. Serious concerns were raised in the 1980s about the impact of acid rain on forests in the United States and Europe. A major scientific study of the problem, the National Acidic Precipitation Assessment Program (NAPAP), was undertaken to test the effects of acidification of soil, water, and forests (NAPAP, 1990). NAPAP and subsequent studies have shown there are more complex relationships between acidic precipitation and plant life than previously thought.

The NAPAP study discovered that, although acid rain did cause the acidification of some lakes in the northeastern United States and Canada, land use practices were implicated as a major contributing cause. Damage to forests was found to be less than originally feared. Forests in both the United States and Europe have made a remarkable recovery in recent years (Kuusela, 1994). Forests in Europe have experienced their most rapid recorded growth during the decade of the 1980s, growing at a rate of 35 percent higher than in earlier decades. This increased growth has occurred throughout Europe, including the regions thought to have been adversely affected by acid rain (Kandler, 1994; Skelly and Innes, 1994).

Textbooks, however, have not been revised to include current scientific theories about this issue. Many discussions of the topic are illustrated with a picture of a dead or dying forest stand, even though further studies have indicated that dieback of some forests is due to causes other than acid rain, particularly climate stress.

Often textbooks do not mention other serious threats from acid precipitation, including health effects. Some texts describe threat to aquatic life but do not explain the poor land management practices that exacerbate the problem.

One text, Lebel's *Environmental Science: How the World Works and Your Place in It*, explains the chemistry of acid rain and includes a discussion of the NAPAP report (Person, 1995). This text, however, is an exception. Holt's *Environmental Science* textbook, published in 1996, discusses impacts of acid precipitation in detail without explaining the natural science or noting that scientific theories have been revised (Arms, 1996). In many biology texts, the discussion of this topic is limited to one or two sentences, often without context. Addison Wesley's *Biology*, for example, states: "Burning coal produces chemicals that form acids when they join with water in the atmosphere. These acids then fall to earth as acid rain, killing trees and fish and disrupting soil chemistry" (Essenfield, Gontang, and Moore, 1994, p. 880). There is no mention here that nitrogen from automobile emissions also contributes to acidification of rainfall and may pose a more serious problem than coal burning plants, because new technology has reduced emissions from that source.

Coverage of acid rain presents an example of a common flaw in the texts. It is inevitable that, particularly in the area of environmental science where the science is new and changing, materials will become dated. Yet most of the discussions in these texts treat the damages presumed to be associated with acid rain as scientific fact. As in other topic areas reviewed by the commission, scientific theorizing on this issue is often presented as settled, although scientists were and are still engaged in speculation and investigation.

This is a troubling tendency. Public policy is, and sometimes must be, made on the basis of incomplete scientific information. However, students should not be misled about the nature of science and the scientific enterprise. In these discussions of science and public policy issues, scientists are often portrayed as omniscient. In the attempt to simplify the discussion (and to fit it into a sidebar), the limits of scientific knowledge are ignored. The implication, therefore, is that scientists know all the answers and that the solutions are obvious. With many—perhaps most—of the environmental issues discussed in these texts, this is not the case. And even on issues where knowledge is reliably certain, science alone cannot provide a remedy. This is one of the most important lessons students should learn from science–technology–society debates. It is critical, therefore, that the role and limitations of science be reflected with more clarity than is now done.

Energy

Coverage of energy and natural resources typifies another common flaw in treatment of environmental issues: Technology is presented simplistically as either the source of all our problems or the cure to all our ills. As *Living Lightly*, a teacher's guide for grades four to six, explains:

> Approximately half of the oil consumed in the United States is used to power automobiles. This consumption takes its toll on people and the environment in many ways. Oil spills degrade coastal ecosystems killing wildlife and destroying fisheries. Burning fuel pollutes the air and contributes to acid rain. Our dependence on foreign oil can also lead to loss of lives and outpouring of billions of dollars in defense. (McGlauflin and O'Connor, 1992, p. 183)

Renewable energy sources, by contrast, are offered as the answer, if only we have sufficient political will.

Without diminishing the challenges of protecting the environment, students should also understand energy in historical context. We have, for example, enjoyed considerable benefits from the availability of relatively cheap, plentiful energy, such as a high standard of living, with accompanying improvements in health and welfare. It will not be easy or without cost to replace the source upon which we rely for 90 percent of our energy needs.

Decisions about energy policy ought to be based on accurate information. One central theme in many textbooks is the rapid depletion of fossil fuels and other natural resources. The *CLASS* Project, a joint effort of the National Wildlife Federation and the California Department of Education, states that "experts believe that by the year 2000, most of the world's oil may be depleted" (California Department of Education, 1992, p. 39). The Globe Fearon's STS text, *Impacts of Technology*, predicts that "supplies of oil will grow scarce in the early 2000s" (Harkness and Helgren, 1993, p. 29). *Environmental Science: Working with the Earth*, published by Wadsworth, states that "at the current rate of consumption, known world oil reserves will last for 42 years" (Miller, 1995, p. 519).

These predictions are based on a misunderstanding of the difference between oil reserves and oil resources. Reserves are the supply of oil exploitable at current prices using available technology. As prices rise, and technology improves, reserves increase. According to the World Resources Institute, "proven recoverable reserves of petroleum rose 60 percent between 1973 and 1993" (World Resources Institute, 1996, p. 275). Reserves increase, despite increasing consumption.

Resources, however, include many sources that are not currently being exploited because it is not cost-effective to do so. Known resources for the production of fossil fuels, including oil, natural gas, coal, tar sands, oil shale, are vast. The extent to which we will continue to rely on fossil fuels for energy use will be determined by technology and the costs of extracting fuels from these resources, compared to the availability and costs of alternative energy sources.

Most textbooks appropriately emphasize the need to develop economically viable renewable technologies such as energy recovery from biomass, wind-energy systems, solar photovoltaic systems, solar thermal energy-conversion systems. Students should also learn that this has been a goal since the early 1970s, but has met with only limited success to date because these technologies are not yet cost-competitive with fossil-fuel technologies. Alternative energy sources will be an important element of future energy policy. Many of the texts reviewed by the commission are disappointing in their coverage of these technologies. Students cannot judge the comparative advantages of solar power if they have had no introduction to technological hurdles, such as problems of transportation and storage, that have to be addressed before solar energy is used widely and reliably.

For example, *Living Lightly on the City* for 4th–6th grades says there are "more than 1 million active solar systems and 250,000 passive solar homes" in the United States (McGlauflin, 1992, p. 183). Indeed, solar technologies for passive space heating and for the production of hot water are being used successfully in many areas. Students will wonder why solar energy is not more widely used if they have no further information about the relative costs and benefits of solar energy.

There are costs to be considered in evaluating the relative advantages of various energy sources. Wind-generators are noisy, and present a threat to birds. Large tracts of land have to be converted for generators, reducing habitats. Lebel's *Environmental Science* is one of few that mentions these aspects; it includes a case study about a California wind field near Alamont Pass that is in a major migration corridor and has created a danger for migrating birds, including golden eagles (Person, 1995). Prentice Hall's *Environmental Science* dismisses this concern about wind turbines by claiming that 300,000 birds were destroyed by the Exxon *Valdez* oil spill (Nebel and Wright, 1998). One commission member noted in response to this argument, "Is it justifiable to kill birds on a continuing basis with wind machines because there was an accident involving fossil fuels?" (p. 595). These considerations make environmental issues a useful tool for eliciting thoughtful analysis, yet too many textbooks miss this opportunity.

Electric-powered vehicles sound promising, but few texts point out the environmental impacts that have to be considered in any comparison.

Wadsworth's *Environmental Science*, for example, challenges the current wisdom that "fuel efficient cars will takes years to develop and will be sluggish, small, and unsafe," and suggests that all major American car companies will have prototype electric cars, some by 1995, that are "extremely quiet, need little maintenance, can accelerate rapidly with adequate power supplies" (Miller, 1995, p. 488). Holt's *Environmental Science* provides a more realistic analysis. It asks students to consider that electric car batteries have to be replaced frequently, and reminds that they contain lead and acid that could leach into groundwater (Arms, 1996). Electricity to run the car has to be generated; the environmental impact of the original source of energy has to be included in a comparative analysis. And because electric cars are expensive, they will likely replace the newest (and therefore cleanest) cars on the roads, rather than older cars that have higher levels of emissions. All these factors need to be considered in assessing relative impacts of various technologies.

Waste Not, Want Not

Waste management issues are also frequent science–technology–society topics and are widely covered in most environmental education resources. Unfortunately, many discussions oversimplify the issues, for example, implying that recycling is a panacea for waste.

Recycling is a cost-effective, efficient use of resources for some, but not all materials. Recycling aluminum cans requires ten percent less energy than producing them from virgin materials. Automobile tires are also successfully recycled, reducing the energy required to produce new tires. Students should understand, though, that recycling is not new. Some form of recycling has been done for over 100 years. For example, since the beginning of the century, waste paper has been recycled to make paper when wood pulp was scarce or too expensive. Recycling of scrap metal began during World War II.

But not all materials are recycled economically or without having an environmental impact. De-inking of newsprint requires using toxics, creating another waste stream that must be safely disposed. Recycling does save trees; however, most new paper is made from trees grown in plantations specifically for that purpose. Recycling other materials, such as glass, can require more energy than is required for production of virgin material, depending on the process used. Students ought to be challenged to look at all these factors in analyzing the role of recycling in waste management policies and, indeed, this type of quantitative analysis provides an excellent opportunity for applying knowledge.

Biology Today, a high school text published by Holt, Rinehart and Winston, states, "Eighty percent of our household trash could be recycled," a statement with which no knowledgeable expert would agree (Goodman, Graham, Emmel, and Schecter, 1991, p. 880).[4] Other texts have useful exercises which engage students in analyzing various products for environmental impact. For example, *Discover the Wonder*, a 5th grade general science text published by ScottForesman, tells students to compare paper and plastic product use based on convenience and reusability (Heil *et al.*, 1994). *Plastics in Our Lives*, by Lawrence Hall of Science, University of California at Berkeley, includes an exercise which asks students to perform a quantitative analysis of the various costs associated with using paper and plastic cups (Chemical Education for Public Understanding Program, 1992). Such an exercise helps students understand that these and other issues are not always as simple as they first appear.

Other waste management issues are covered in a facile manner. Waste incinerators, used widely in Europe to convert trash to energy, are treated as a major source of concern. Waste incinerators are used as the introductory exercise in Globe Fearon's science–technology–society textbook for middle school, *Impacts of Technology* (Harkness and Helgren, 1993). Students are instructed to read an article about a community's debate over a waste incinerator and are asked to weigh evidence given for and against the construction. The only scientific evidence offered is by the local pollution control agency, which has tested and found that ash is not a hazardous waste. Environmentalists and others in the community have expressed concern. The students are asked to decide their position on the issue. However, all the activities following their analysis include methods the students can take to lobby against the incinerator. Regardless of one's position on the relative safety of waste incineration, we can ask if we are effectively teaching decisionmaking skills if we presuppose the "right" answer.[5]

Forests

Forest management and the disappearance of rain forests is a common theme in environmental materials and STS exercises. Coverage of forestry issues varies considerably in accuracy among the materials reviewed. Estimated rates of deforestation, for example, differ widely from one text to

---

[4] Keep America Beautiful's comprehensive study, *The Role of Recycling in Integrated Solid Waste Management to the Year 2000*, predicts that the most optimistic projection for recovery for recycling and composting is 35 percent.

[5] This is a surprisingly common flaw in many of the texts reviewed by the commission. The text itself will accurately discuss the various positions on a controversial topic. The questions or activities accompanying the unit, however, will assume one "correct" position, and students are instructed to defend it. For many issues, even environmentalists disagree about the best approach. What if the student disagrees with the answer chosen?

another, without citation or reference to the source of the data. The causes of rapid deforestation, where it is a problem, are usually not defined.

Discussions of temperate forests often do not note the remarkable recovery that has occurred in the United States and Europe in recent decades. High efficiency farming, which is depicted in other chapters as the source of many environmental problems, has permitted large tracts of land, particularly in the eastern United States, to revert to forests. Most commercial logging in the United States occurs in tree plantations or second-growth forests and almost all old-growth stands are now under protection in national parks or protected by policies not to harvest old-growth remaining in public forests. Similar recovery of temperate forests has occurred in most western nations. Globally, the FAO estimates that there has been a net increase in temperate forests from 1980 to 1990 (U.N. Food and Agriculture Organization, 1992).

Causes for deforestation are often poorly identified. *Living Lightly on the Planet*, for example, tells students, "rich nations around the world enter the lumbering business in the rain forest and lay waste to vast areas" (O'Connor, 1995, p. 22). Clearing land for agriculture, not commercial logging, is the major cause for deforestation in many developing countries. Prentice Hall's *Exploring Life Science*, includes a sidebar that informs students "many people in the United States as well as other countries are trying to find ways to save the rain forest," and asks, "Do you think the rain forest should be saved?" (Maton, 1995, p. 733). What information do the students have that would allow them to address such a question in a thoughtful way? The students have learned nothing in the chapter about the economic incentives in some developing countries that have led to accelerated clearing of forests, such as insecure tenure to land or government subsidies for forest conversion. A better presentation of this issue is found in Prentice Hall's *Environmental Science*, which includes a discussion of the economic pressures that has led to severe forest loss in some developing countries. Students are asked to consider "What incentives and assistance could the United States offer Brazil or Guyana to keep their tropical rain forests from further harmful development?" (p. 506). This approach calls upon students to move beyond expressing opinions to thinking critically about real problems and possible solutions.

Ecology

Central to the study of the environment is an understanding of ecology and biodiversity. The commission found some of the best materials on ecology and biodiversity in the texts reviewed. *Biological Science: An Ecological Approach*, published by Kendall/Hunt, presents college-level ecology at the high school level (Milani *et al.*, 1995). *Eco-Inquiry*, developed by the Institute of Ecosystem Studies and published in 1994 by Kendall/Hunt, takes a few fundamental ecological problems and makes them understandable and interesting to young students. This text, unlike most others reviewed by the commission, engages students in the process of scientific discovery and helps them to learn about the methods underlying the work.[6]

However, there are certain problem areas. For example, rates of species extinction are often stated as fact, without acknowledging the uncertainties surrounding the current state of scientific knowledge with respect to species extinction (e.g., Maton, p. 101). *Environmental Science*, published by Holt, Rinehart and Winston, provides a dated estimate of the number of species (Arms, 1996). The most commonly accepted figure, not counting microorganisms, is around 10 million, not between 10 and 100 million (Dobson, 1996; Reaka-Kudla, Wilson, and Wilson, 1997). The same text explains only the role of habitat destruction in species extinctions, when other causes, such as introduced species and harvest, should be covered as well.

Population and Hunger

Population and hunger are also issues in which oversimplified explanations can be misleading. The field of "population studies" encompasses an array of separate and distinct disciplines, including biology, applied mathematics, economics, sociology, and history. Because these questions bring together the natural and social sciences, they can be particularly challenging. As the commission notes:

> Whereas relationships examined in the physical sciences are typically universal and immutable (e.g., the conservation of energy), relationships in the social sciences depend critically upon human behavior, which can differ fundamentally in separate settings and over time. If the distinction between natural science and social science components within population studies is not effectively outlined and explained, there is the risk that students will come away applying inappropriate scientific paradigms to modern population questions. (ICEE, 1997, p. 23)

In some of these texts, "overpopulation" is asserted as a major cause of environmental problems. Addison-Wesley's environmental science textbook, for example, begins its discussion of the topic with the statement: "Scientists argue that overpopulation threatens the continued existence of

---

[6] The commission noted (as have others) that science textbooks far too often present a dry set of facts. Students are engaged in little of the *process* of science: observation, analysis, hypothesis, prediction, and test of prediction, the results of which are then incorporated into the body of science or abandoned.

humans on earth" (Bernstein, 1996, p. 204). In fact, "overpopulation" is not a scientific concept used in either demography or population studies because of its ambiguity. Many of the problems raised in discussions of population stem from poverty rather than population growth.

Population trends are usually presented as fixed. Yet demographers face considerable uncertainties in their attempt to project future population trends. Only a few materials under review mention that the population growth rate is declining throughout the world, even in developing countries, with the exception of sub-Saharan Africa. A number of texts, such as LeBel's environmental science textbook (p. 253), have students calculate the "doubling time" of population as a math exercise. This exercise, though, assumes that population will grow at a constant rate for decades or generations. In historical experience, this has almost never happened. Demographers must constantly revise their estimates to reflect demographic transitions as countries develop. As a country or region's economy improves, human fertility levels eventually tend to decline. In Europe, North America, and parts of the Caribbean and high-income East Asia, fertility levels today (if maintained indefinitely) are lower than would be necessary for long-term net population replacement.

While population growth has presented unparalleled pressures on natural resources, discussions of the issues often neglect to mention why global population has increased so rapidly in recent history. Population is growing around the world because of worldwide improvements in health, not because of fertility trends. Between 1950 and 1990, fertility levels declined in almost every region of the world. With the exception of a few current problem areas in the former Soviet Union and sub-Saharan Africa, longevity is increasing and infant mortality is decreasing. Global life expectancy has more than doubled since 1900, and the infant mortality rate has declined by over 50 percent between the early 1950s and the early 1990s.

Famine is simplistically attributed to overpopulation in many of the materials, without an explanation of the role of the governments involved. Prentice Hall's *Biology: The Study of Life* illustrates its section on population with a picture of swollen-bellied child, victim of famine in Africa (Schraer and Stolze, 1991, p. 865). The text states famines result because food production has not increased as fast as population. Students are asked to write a letter to the editor discussing "how to end world hunger." To write such a letter, though, students would need a better explanation of the causes of world hunger, which has less to do with population growth than with political conditions. Recent famines in Africa were not caused or exacerbated by overpopulation, adverse weather, or crop failure. Specific political factors, such as civil wars in which food has been used as a weapon and which interrupted food distribution and agriculture. Since the goal of environmental education is to help students understand how human actions affect the environment, it would seem particularly appropriate to help them understand the political context of modern famines.

Poverty, not simply population growth, is responsible for many environmental problems faced by underdeveloped nations. Deforestation in many underdeveloped countries continues because people have access to no fuel other than wood. Insecure property rights in many nations have led to careless land management, deforestation, and soil erosion. The relationship between economic development, political stability, and a nation's willingness or ability to address environmental concerns is often not well explained. Studying population growth as a simple numerical equation (more people equals less food) does little to help students comprehend the serious environmental problems faced by developing nations.

One text, Globe Fearon's, *Environmental Science: Changing Populations*, provides a thoughtful introduction to current scientific and economic thinking on population change (Falk, 1995). The text correctly informs the student, for example, of the problems inherent in making long-term population predictions. A number of activities require students to consider the tradeoffs and interactions involved in both environmental policies and economic development or population change.

Understanding Risk

Difficult decisions, such as how to protect endangered species or what to do about abandoned hazardous waste sites, capture students' interest. Many of these issues, moreover, have immediate impact on students' daily lives. Students need some knowledge of risk analysis to prepare them to participate as policy makers and informed citizens in public decisionmaking.

Risk analysis is the process by which scientific information is distilled so that it becomes useful for making decisions. Few of the materials reviewed convey key concepts that would help students understand the nature of risk. For example, central to risk analysis is the dose–response (or exposure–effect) relationship. The likelihood of harm, or the severity of harm, from any substance rises when the amount of exposure increases. Virtually any substance or activity can produce adverse effects if the exposure reaches high enough levels.

Instead of introducing students to this critical concept, the materials under review focus on enumerating hazards without discussing likelihood or size of the danger. For example, the Prentice Hall's environmental

science text presents a table of the known health effects, including muta-
genicity, teratogenicity and carcinogenicity, of synthetic organic chemicals
(Nebel and Wright, 1998, Table 11-1, p. 349). Each chemical has a box that
is checked if exposure to that chemical presents a health threat. The table
contains no information about the dose (or exposure) that might be necessary to achieve this harm. Students are not told, for example, that aspirin
would be considered a health hazard in this analysis.

Considerable attention is paid to the harm chemicals can cause,
without addressing the amount of exposure necessary to cause the harm.
Biology: The Dynamics of Life, a high school text published by Glencoe,
includes as a "Thinking Lab" a discussion of a report by an environmental
group that examined fruits and vegetables for pesticide residue (Biggs,
Kapicka, and Lundgren, 1995). Most of the fruits and vegetables tested were
found to have some residue. Students were asked to conclude which of the
fruits and vegetables (including most common fruits and vegetables such as
bananas, apples, celery, and broccoli) shown were most likely to present a
danger. The assumption here is a prime example of focusing on a potential
hazard and ignoring the size of the risk. A recent National Research Council
report found that toxic chemicals occuring naturally in foods may pose a
greater health threat than pesticide residues, and that the greatest threat to
human health is a poor diet, particularly a diet poor in fruits and vegetables (National Research Council, 1996). It would be unfortunate if this
sidebar persuades students to reduce the amount of fruits and vegetables
in their diet.

Discussions of pesticides rarely consider the relative costs and effects
of alternatives to using pesticides. A few well-known cases, such as DDT,
are used to predict that other (or all) human-made compounds, even where
the risk is not yet known or is not indicated by epidemiological evidence,
are dangerous.

The Wadsworth environmental science textbook is one of the few that
includes comparative risk information, listing high risk health problems,
such as indoor air pollution, and pollutants in water (Miller, 1995, Table
8-1, p. 205). However, students are never told that these risks are small
compared to the health risks from smoking or car accidents.

Risk Comparison, published by Addison-Wesley and developed by
the Chemical Education for Public Understanding Program, Lawrence Hall
of Science, University of California, Berkeley, provides an intelligent and
understandable introduction to risk (CEPUP, 1900). Students are introduced to the nature of risks through exercises that ask them to consider
some of the risks they take in their lives, such as vaccinations. Some exercises, however, rely more on students' perceptions to assess relative risks
than on actual data indicating probability of harm.

Environmental Decisionmaking: Trade-offs

Over 25 years of experience in managing environmental risks have demonstrated that trade-offs are inevitable. Environmental problems often are not easy to mitigate and solutions are often expensive. Trade-offs arise when an action taken to reduce a risk actually creates risks of its own. For example, to comply with regulatory requirements for fuel economy, mandated under the Clean Air Act, automobile manufacturers began to build lighter and smaller cars. As a result, fatalities and serious injuries from automobile accidents increased. To improve safety, air bags were installed in most cars and were mandated for all cars manufactured after 1996. Air bags, however, have been responsible for a number of serious injuries and fatalities in low speed crashes, particularly in small children and elderly women.

Many environmental teaching materials fail to convey the trade-off of one risk for another that pervades environmental management efforts. Instead, in descriptions of actions to reduce environmental damage, textbooks often suggest that the solution is obvious and that people are simply not taking action. The commission stated, "this has the unfortunate result of making adults look either inept or irresponsible because they refuse to take the actions necessary to save the environment or to protect public health" (ICEE, p. 39). Students should understand that virtually every environmental management question presents difficult scientific issues, uncertainties, competing social values, and trade-offs.

For example, LeBel's environmental science textbook titles its chapter about alternatives to pesticides, "Integrated Pest Management— A Better Way to Control Pests" (p. 287). The discussion on agricultural pesticide use suggests that biological controls, such as predator species for pests, are a ready solution to the problem of "toxic" pesticides. However, the presentation does not cover the risks of using biological controls. Predator species may control pests in some circumstances, but if the predator is not effective (and only a small fraction is effective) a farmer may suffer significant crop losses. Another potential risk may occur from introducing a predator species that becomes a pest in its own right and damages other wildlife in the area. Such has been the case with the small Indian mongoose, which caused the extinction of some reptiles in the West Indies.

Virtually every environmental issue presents an opportunity to engage students in analyzing trade-offs. If the trade-offs are not considered, students may develop simplistic and misleading perceptions of issues. For example, high per capita energy use in affluent nations permits a higher standard of living in terms of health, comfort, and leisure. Where should sacrifices be made if we want to reduce our level of energy use? As noted previously, topics such as recycling or alternative sources of energy are best understood if students have some understanding of the costs and benefits involved.

A few teaching materials do a good job of introducing students to the trade-offs inherent in environmental decisions. *Plastics in Our Lives*, from Lawrence Hall of Science, as previously noted, has students consider alternative ways to carry groceries: paper bags, plastic bags, and mesh or cloth bags. Students learn about the different energy requirements for making each type of bag, how each is recycled, and the durability and capacity of the bags. They create a matrix of the scientific evidence and then discuss which factors might be important and deserve greater weight in a decision. From this simple exercise students learn that few decisions are straightforward, that quantitative scientific information is necessary, and that social values are key to making good decisions.

Critical Thinking?

Environmental topics can be used to challenge students to think critically about controversial issues. However, texts too often fail to provide enough information to allow students to understand and discuss environmental controversies thoughtfully.

Science–technology–society and environmental science courses tend to be more prevalent at the middle school level. The STS series of texts published by Globe Fearon, for example, targets middle school classes. So are environmental science textbooks, or they are aimed at students who will not be taking biology, chemistry, or physics.[7] Science–technology–society sidebars are more prevalent in middle school textbooks than in upper level textbooks. This appears to reflect trends in educational theory and practice: It is easier to study multidisciplinary topics at the middle school level because the disciplines are not so clearly delineated as they are in secondary courses. This trend is unfortunate. In these lower level texts there is a tendency to oversimplify what are usually complex issues. The National Research Council, in its study of biology education, finds:

> We are concerned that courses offered as "science–technology–society" (STS) do not follow a study of the basic sciences. Instead, they typically replace basic-science courses, and that results in both a dilution of fundamental knowledge of basic sciences and a lack of the scientific breadth need to study interdisciplinary topics more than superficially. (NRC, 1990, p. 57)

---

[7] See, for example, National Science Teachers Association's *Science Teacher* 64 (December 1997), 11. Advertisement for LeBel's environmental science text claims it is "designed to introduce STS and environmental issues. Ideal for students not taking chemistry or physics." Also see, Singletary, T. (1992). Case studies of selected high school environmental education classes. *Journal of Environmental Education 23* (4): 35–40, 48. Teachers interviewed reported that environmental science courses were typically offered as a general science course for students who did not plan to take more science classes.

Environmental science and STS offer a wonderful opportunity for multidisciplinary investigation of real problems. Students who have not yet begun to study science and other disciplines in depth will have only limited background to inform their investigation of these issues. The National Research Council concludes, "the contribution of science to the solution of social problems can be understood only when there is considerable understanding of science itself."

If we are to succeed in creating a generation of well-informed citizens, we must prepare students to make responsible environmental decisions. This requires that students understand the science involved, the comparative risks, the economic and social trade-offs inherent in any environmental policy.

The goal of science–technology–society is that students will understand the relevance of science-based issues to their everyday lives. But many science-technology-society exercises included in textbooks fail to encourage student to use scientific skills in analyzing the issue. Many of these issues have been studied by thoughtful people for years but uncertainties and disagreements remain about the scope of the problem and the most effective solutions. STS issues, though, are often presented in a sidebar, with a brief description of the problem and questions that elicit opinions from students about appropriate solutions. Rarely are references provided to the source from which the data is derived, or references for further study, nor is the student encouraged to test the validity of the information or the assumptions made in presenting the dilemma. At best the student can only offer a glib response based on incomplete information and faith in the validity of the data used as evidence.

For example, following is an entire STS sidebar in LeBel's Environmental Science:

> The consensus in the scientific community is that assessing environmental health risks is imprecise. The "baby boomer" generation for example suffers a cancer rate three times that of their grandparents. However the EPA continues to calculate and regulate industry according to 'acceptable' levels of human exposure to toxins. Do you think there is "acceptable" levels of risk for humans and other organisms? (Person, 1995, p. 139).

A scientist would immediately ask many questions. First, is the data accurate? What is the source of the data?[8] If the data is accurate, what does it

---

[8] It is exceedingly rare for a textbook to furnish citations for even the most astounding statistics. For example, Wadsworth' environmental science textbooks includes a thought-provoking "fact" at the bottom of each page, without reference or explanation. On page 291, the text asks, "How many cancer deaths in the United States are caused by exposure to

explain, and not explain? For example, cancer is a disease of old age. Does an increase in the rate of cancer have any relation to increased longevity? Are people living long enough to die of cancer, rather than from tuberculosis, influenza, and other insidious diseases which were once the major causes of mortality? Are all cancers on the rise? What percentage of the increase could be due to lifestyle factors, such as smoking and lung cancer? What evidence is there that low-levels of toxins are correlated to cancer? As noted above, evidence indicates that lifestyle factors are a greater threat to health than residue toxins on food. No reference is made to the fact that not all toxins are human-made, or that naturally-occurring toxins as well as human-made toxins may be carcinogenic, nor that scientists and government agencies have studied these questions for many years without resolution.

Such an exercise does not help students analyze issues thoughtfully. This is a serious failing. There are long-term consequences for rational decisionmaking when citizens and policymakers are prey to misinformation and susceptible to those who would capitalize on scientific uncertainties for purposes of persuasion. Allocating large amounts of scarce resources to chase "phantom" risks means that those resources are unavailable to address other risks that may have more serious impact on human health.

Students would be much better prepared to participate in debates about public policy if they are trained in methods of scientific inquiry. STS exercises offer an excellent opportunity to challenge students to analyze data critically. The National Science Standards and Benchmarks for Scientific Literacy call for all students to understand concepts such as probability, random sampling, and the difference between correlation and causation. This latter concept is particularly important. One of the most common misuses of statistics in public policy is the confusion of correlation with causation. The previous example shows a typical way in which these concepts are confused. Simply observing that two phenomena occur at roughly the same time does not demonstrate, without other evidence, that one phenomenon is caused by the other. Examples of this error abound in the media and everyday life, and form the basis for many interesting conspiracy theories. Unfortunately, this error is not infrequent in discussions of environmental issues, as the previous example demonstrates. Decisionmaking exercises in science textbooks should be used to help students learn to *critically* assess the validity of statistical evidence.

---

pesticide residues in food? Answer: 4,000 to 20,000." There is no explanation of what this means, where the data come from, or even an acknowledgement that there is much that is not yet understood in this area. This does not present a good example for young scientists or informed citizens.

RECOMMENDATIONS

It is clear from the commission's review of commonly used science textbooks that science–technology–society and environmental issues are not well integrated into the discussion of science concepts in the curriculum. These issues are usually "add-ons," sidebars labeled "science–technology–society" or "thinking labs" and are not edited with the same attention given to the remainder of the text. Presenting complex issues in such a superficial way inevitably leads to misleading oversimplifications that do not facilitate critical thinking. Teaching materials that encourage students to defend uninformed opinions are not educational.

The commission has made several recommendations that it believes will help to improve teaching in this area:

Experts in environmental science should review textbooks and other educational materials on environmental issues. Environmental issues integrate a number of scientific disciplines. Scientific reviewers from all relevant disciplines should review textbooks to help insure their accuracy, and publishers should take their reviews seriously. In view of the importance of providing a quality education for the next generation, professional scientists, particularly research scientists, have a duty to become more involved in reviewing resources for the K–12 classroom than they are now.

Publishers should also make sure that qualified experts review textbooks after they have been edited and that sidebars, science–technology–society exercises, and other ancillary materials receive the same level of review.

Discussion of environmental issues should help students understand the underlying scientific concepts and should be linked to the science curriculum. More substantive analysis of environmental issues should be included in middle school and upper-level science texts. Science activities should be consistent with the criteria for scientific inquiry as set out in the National Science Education Standards, and should motivate students to acquire a deeper understanding of the science.

Environmental science high school textbooks should be revised to introduce students to environmental science in the same way that biology or chemistry textbooks introduce high school students to those sciences. These textbooks should provide a rigorous, substantive introduction to environmental science rather than, as many now do, simply a catalog of environmental facts and issues.

Schools should consider adding a multidisciplinary capstone course on science–technology–society issues, including environmental topics. While it is useful to introduce students in lower level classes to these issues, students will not be able to study them with the same depth of understanding as upper-level students who have had more background in both the natural and social sciences. The National Research Council proposed that courses that integrate science and social issues be taught as an upper-level capstone course. An upper-level course would permit students to apply knowledge acquired in biology, chemistry, and physics courses, with input from social studies, economics, and political science courses for an indepth and thoughtful investigation of an environmental issue. This approach is ideal, but it may be difficult to incorporate into the curriculum. However, because of the interest that many students have in the environment, an upper-level capstone course would likely be well subscribed and could possibly motivate students to study science in college.

CONCLUSION

Ours is increasingly a technological world. Students in classrooms today can scarcely imagine a world without computers, televisions, or satellite communications, yet these technologies were unknown a hundred years ago. In 1943, Thomas Watson, chairman of IBM, could make the statement, "I think there is a world market for about five computers." There are limits to our prescience. We will not be able to teach students today everything they will need to know in the next millenium.

But we can prepare them. There are few public policy debates, particularly about health and natural resources, that do not resort to evidence derived from the sciences—epidemiology, climatology, or ecology, for example. Technology is a pervasive part of our lives, but technological advances often entail unexpected costs and risks in addition to benefits. A clear grasp of the interrelationships between science, technology, and society is essential to basic science literacy. Students need to be able to critically evaluate evidence, to distinguish between scientific knowledge and knowledge obtained by other means, and to understand that the scientific enterprise must be governed by societal values.

Environmental-related STS issues have risen to the top of the public agenda, and will continue to dominate public debate for the foreseeable future. Although particular environmental issues may change, the skills and knowledge students build in the classroom today will prepare them to face the challenges of tomorrow. The current approach of adding environmental study to the curriculum as supplemental activities or sidebar exercises is inadequate, because it encourages students to base opinions on superficial study rather than true scientific inquiry. Educators should give further thought to incorporating study of the environment as a rigorous, substantive, and integral element of the science curriculum.

REFERENCES

Arms, K. (1996). *Environmental Science*, Holt, Rinehart and Winston, Boston.

Bernstein, L., Winkler, A., and Zierdt-Warshaw, L. (1996). *Environmental Science: Ecology and Human Impact*, Addison-Wesley, Menlo Park, CA.

Biggs, A., Kapicka, C., and Lundgren, L. (1995). *Biology: The Dynamics of Life*, Glencoe/McGraw-Hill, New York.

California CLASS Project (1992). California Department of Education/National Wildlife Federation, Sacramento.

Chemical Education for Public Understanding Program (1992). *Plastics in Our Lives*, Addison-Wesley, Menlo Park, CA.

Chiang-Soong, B. (1993). STS in most frequently used textbooks in U.S. secondary schools. In Yager, R. (ed.), *What Research Says to the Science Teacher. Vol. VII*, National Science Teachers Association, Washington, D.C., pp. 43–47.

Disinger, J. (1986). Locating the "E" in S/T/S. *ERIC Information Bulletin* 3:1–3.

DiSpezio, M., Lisowski, M., Skoog, G., Linner-Luebe, M., and Sparks, B. (1995). *Science Insights*, Addison-Wesley, Menlo Park, CA.

Dobson, A. P. (1996). *Conservation and Biodiversity*, Scientific American Library, New York.

Essenfield, B., Gontang, C., and Moore, R. (1994). *Biology*, Addison-Wesley, Menlo Park, CA.

Falk, D. (ed.) (1995). *Environmental Science: Changing Populations*, Globe Fearon, Paramus, NJ.

Gambro, J., and Switzky, H. A. (1996). National survey of high school students' environmental knowledge. *Journal of Environmental Education* 27(3):28–33.

Goodman, H., Shecter, Y., Graham, L., and Emmel, T. (1991). *Biology Today*, Holt, Rinehart and Winston, Chicago.

Harkness, J., and Helgren, D. (1993). *Technology and Society: Impacts of Technology*, Globe Fearon, Paramus, NJ.

Hausbeck, K., Milbrath, L., and Enright, S. M. (1992). Environmental knowledge awareness and concern among 11th-grade students: New York state. *Journal of Environmental Education* 24(1):27–34.

Heil, D. (1994). *Discover the Wonder-Grade 5*, ScottForesman, Oakland.

Hurd, D., Matthias, G., Johnson, S., Snyder, E., and Wright, J. (1992). *Science: A Voyage of Exploration*, Prentice Hall, Englewood Cliffs, NJ.

IEA TIMSS International Mathematics and Science Study. Performance on Items Within Each Science Content Area, http://www.ustimss.msu.edu/frame.htm.

Independent Commission on Environmental Education. (1997). *Are We Building Environmental Literacy?* George C. Marshall Institute, Washington, D.C.

Kandler, O. (1993). The air pollution/forest decline connection: The "Waldsterben" theory refuted. *Unisylva* 44:173.

Kempton, W., Boster, J., and Hartley, J. (1996). *Environmental Values in American Culture*, The MIT Press, Cambridge, MA.

Knapp, C. (1996). A response to Bora Simmons's recent president's message. *Environmental Communicator* 12:21–22.

Kuusela, K. (1994). *Forest Resources in Europe*, Cambridge University Press, Cambridge.

Maton, A. (1995). *Prentice Hall Science: Exploring Life Science*, Prentice Hall, Englewood Cliffs, NJ.

Maton, A. (1994). Prentice Hall *Science: Ecology: Earth's Living Resources*, Prentice Hall, Englewood Cliffs, NJ.

McGlauflin, K., and O'Connor, M. (1992). *Living Lightly in the City*, Schlitz Audubon Center, Milwaukee.

Milani, J., Leonard, W., Manney, T., Rainis, K., Uno, G., and Winternitz, K. (1992). *Biological Science: An Ecological Approach*, Kendall/Hunt, Dubuque, IA.

Miller, G. T. (1995). *Environmental Science: Working with the Earth*, Wadsworth Publishing Co., Belmont, CA.

McFadden, C., and Yager, R. (1993). *Science Plus: Technology and Society*, Holt, Rinehart and Winston, Chicago.

National Research Council (1996). *Carcinogens and Anticarcinogens in the Human Diet: A Comparison of Naturally Occurring and Synthetic Substances*, National Academy Press, Washington.

National Research Council (1990). *Fulfilling the Promise: Biology Education in the Nation's Schools*, National Academy Press, Washington.

Nebel, B., and Wright, R. (1998). *Environmental Science*, Prentice Hall, Upper Saddle River, NJ.

O'Connor, M. (1995). *Living Lightly on the Planet*, Schlitz Audubon Center, Milwaukee, WI.

Person, J. (1995). *Environmental Science: How the World Works and Your Place In It*, J. M. Lebel, New York.

Reaka-Kudla, M., Wilson, D., and Wilson, E. O. (eds.) (1997). *Biodiversity II: Understanding and Protecting our Biological Resources*, Joseph Henry Press, Washington.

The Roper Organization (1992). Teen America's Environmental GPA. A survey commissioned by S.C. Johnson, Inc.

Rubba, P., and Wiesenmayer, R. (1998). Goals and competencies for precollege STS education: Recommendations based upon recent literature in environmental education. In Hungerford, H., Bluhm, W., Volk, T., and Ramsey, J. (eds.), *Essential Readings in Environmental Education*, Stipes Publishing L.L.C., Champaign IL, pp. 327–336.

Schraer, W., and Stolze, H. (1991). *Biology: The Study of Life*, Prentice Hall, Englewood Cliffs.

National Acid Precipitation Assessment Program (NAPAP) (1990). *The Causes and Effects of Acidic Deposition, Vol. IV*, Government Printing Office, Washington, D.C.

Skelly, J., and Innes, J. (1994). Waldsterben in the forest of central Europe and eastern North America: Fantasy or Reality? *Plant Disease* 78(11):1021–1032.

Stapp, W. (1969). Environmental encounters. *Environmental Education* 1(1):30–31.

United Nations Economic Commission for Europe UN-ECE/Food and Agriculture Organization of the United Nations FAO (1992). *The Forest Resources of the Temperate Zone: Main Findings of the UN-ECE/FAO 1990 Forest Resource Assessment*, Report No. E. 92. II. E. 27, United National, New York.

World Resources Institute (1996). *World Resources 1996–1997*, Oxford University Press, New York.

Yager, R. (ed.) (1993). *What Research Says to the Science Teacher Vol. VII: The Science Technology Society Movement*, National Science Teachers Association, Washington, D.C.

CHAPTER 7

Marginalization of Technology within the STS Movement in American K–12 Education

Dennis W. Cheek

A MATTER OF DEFINITION

We live in a pervasive, technological world. Technologies of all types are part of the fabric of everyday life. They extend human capabilities, aid in the prevention of disease, facilitate human interaction, structure commerce, and provide endless hours of entertainment. They also provide means and methods for social control and realization of the worst in human behaviors. Yet technology as a major arena of human activity and engagement scarcely enters the consciousness of most Americans, despite the fact we interact with technological systems and artifacts virtually all of our waking moments and even when we sleep (McGinn, 1991; Melzer, Weinberger, and Zimman, 1993; Latour, 1996).

The STS movement in the United States within K–12 education recognizes the need to "develop scientifically literate individuals who under-

Dennis W. Cheek, Rhode Island Department of Education, Providence, RI 02903.

Science, Technology, and Society: A Sourcebook on Research and Practice, edited by Kumar and Chubin, Kluwer Academic / Plenum Publishers, New York, 2000.

stand how science, technology, and society influence one another and who are able to use this knowledge in their everyday decision-making," as the 1982 position paper on STS from the National Science Teachers Association framed it (National Science Teachers Association, 1982). Technology has been viewed from the outset of the movement in American K–12 education as one of three critical legs of the STS stool.

Yet considerable confusion arises in K–12 education circles when one mentions the word "technology." To many K–12 teachers and administrators, technology refers absolutely and circumspectly to computers, computer networks, software, and related devices that are part of the Information Age. Most schools, by this definition, are not only explicitly aware of technology but they consciously use it on a daily basis and students are frequently instructed in its use.

Technology, as we will employ the term in this paper, is of much older vintage than modern information technologies. Although technology includes computers and related devices, it also embraces the entire human-constructed world of artifacts and systems (Webster, 1991; Volti, 1995). Professor Stephen J. Kline (1985) suggested technology is a complex set of concepts, artifacts, and systems, that can be discussed in four major ways:

1. As artifacts or hardware, e.g., pencils, microscopes, antiballistic missiles
2. As sociotechnical systems of production, e.g., an automobile assembly line
3. As technique or methodology, e.g., the skills, knowledge, and general know-how to rebuild an engine or to engage in oil painting
4. As sociotechnical systems of use, e.g., an airplane presupposes a much wider system of rules and regulations, licenses and trained pilots, passengers and cargo, maintenance, airports, manufacturing facilities, air control, etc.

Technology is the oldest of human endeavors. Early tools, art, production of clothing, human language, and symbolic communication all are examples of technologies in use since the dawn of time. Technology predates by thousands of years the advent of other fields of human endeavor such as science, history, and the social sciences. Technologies evolve in response to changing human needs or environments. Developments in technologies are influenced by a variety of factors including available materials, time, creativity, market demand, and prevalent ideas and beliefs within human cultures in terms of religion, philosophy, and social mores. For these reasons, human beings have often created very different technologies to meet the same basic human needs. Ancient cultures that depended on rivers for their existence, for example, evolved a variety of river artifacts and systems to aid transportation and commerce, exploitation of the river's resources, and management of the river's course (McAdams, 1996; Westrum, 1991).

All technologies embody the explicit and implicit values of their creators (Ellul, 1990; Green, Owen, and Pain, 1993; Morgall, 1993). A chair, for example, in a modern manufacturing plant embodies the concept of "normality" or "average" in terms of its dimensions. It presumes certain things about the unknown user including the length of their limbs, the amount of sustained time they might spend in the chair, and varied uses for the chair. It also reflects views of its creators with regard to style, color, and "feel." A handmade chair created by a colonial craftsman, on the other hand, while often more individually tailored for a particular user, also unavoidably embodies certain values of its maker (Pound, 1989).

Another key concept for all technologies is the idea of "trade-offs." Each technological artifact, system, or methodology conveys certain benefits while imposing certain burdens or costs associated with its use or implementation. For example, a statewide testing system enables central policymakers, the public, and other interested individuals and organizations to get a read on how well the system is doing relative to certain valued ends as measured by the testing instruments. On the other hand, such a system also involves direct financial and other costs because of its creation, dissemination, administration, and its reporting of the results. Some users benefit from the technology, others suffer at its hand, while still others neither benefit nor suffer (Wenk, 1995; Winner, 1986).

Every technology also results in unanticipated consequences for users and others affected by it (Rothenburg, 1993; Sarewitz, 1996). These consequences cannot be forecast in advance by the designers of the technology but come to the fore as particular technologies are implemented in situations not within the purview of the original design work (MacKenzie, 1996). For example, the first paved roads in American cities came into being because of the huge amounts of horse droppings that had to be collected from city thoroughfares and because carriages were getting stuck on muddy avenues. This network of paved streets became an ideal means of conveyance for the first "horseless carriages" and promoted their rapid adoption by affluent city dwellers. Developers of the "peaceful uses of atomic energy" in the United States in the fifties did not foresee the present problems of low level radioactive waste disposal, nuclear power plant failures and decommissioning, and public opposition to expansion of power plant sites (Bauer, 1997; Marcus and Segal, 1989; Segal, 1994).

Within modern science and technology, the boundary between these two fields of endeavor is becoming increasingly blurred. Largescale research and development projects, such as the Human Genome Project, global change research, or Intelligent Transportation Systems, involve thousands of scientists, engineers, and technicians. At any given point, on any given day, a freeze frame of activity would lead to an unresolvable debate among purists as to whether science or technology were being utilized. Despite these convergences, science is fundamentally engaged in explaining the workings of the natural world while technology is fundamentally concerned with taking raw materials in the natural world and blending them with human expertise and creativity to create products, goods, and services that meet human needs. Many people, especially those in science education, erroneously define technology as "applied science." While in some specific cases such a designation holds true, in many other instances technology is employed with little or any explicit use of scientific knowledge or understanding (Marcus and Segal, 1989; Segal, 1994; Mitcham, 1994).

Design and design constraints play a major role in technology development and evolution. The very tools used in the design process (e.g., CAD, CAM) are themselves technologies, and therefore subject to trade-offs inherent in all technologies. Each tool has certain advantages and disadvantages. The goal of all design activities is to optimize the design. This value-laden process is vital to market success, health and safety, and customer satisfaction. Optimization involves a complex balancing of competing desires from those commissioning the design work. The optimization process also requires a seemingly endless round of tests, redesigns, and retests, until a desired balance is achieved in the overall design and its performance. There is rarely, if ever, an example of a design process where no failures or redesigns were called for on the basis of initial test results (Billington, 1996; DeVries, Cross, and Grant, 1993).

Not all technologies are developed explicitly with human "needs" in view. Market pull is often a key factor in the development of many technologies, but companies also use market "push" to bring new products to market. Sometimes a new product is developed on the basis of an individual's or group's perception of a "neat idea." The company conducts some preliminary market analysis and then advertises to create market demand for the new technology. You see this market push most clear in the world of children's toys where each season brings a new raft of consumer products that a child "just has to have." (Pursell, 1995; Nye, 1994).

Appropriate uses for varied technologies and the impact of technological systems have to be constantly monitored by an alert citizenry to ensure that democratic ideals are upheld and that values implicit in civic life are promoted or at least not stifled by technological advances (Sclove, 1995; Simpson, 1995). This includes attention to the differential impacts of various technologies on subcultures or on gender (Wajcman, 1991). The historical evolution of a particular technology frequently serves as a useful analogue or predictor for the potential future impact of a technology through changing culture, modifying social behavior, or having an impact on political and religious life (Stevens, 1995).

TECHNOLOGICAL LITERACY

Clearly, technology of the kind discussed in this chapter, embraces a wide variety of concepts and principles. Current educational parlance suggests that we need to help all students attain "technological literacy" or what sometimes has alternatively been called "technological fluency." The first national Technological Literacy Conference was held in Baltimore, Maryland in 1986. A year later, the second conference sought to better define what "technological literacy" was desired by students leaving secondary schools and students graduating from colleges and universities (Waks, 1987). An Advisory Committee to the National Science Foundation (1996) issued a report on undergraduate education in science, mathematics, engineering, and technology that echoed this earlier call for technological literacy. It noted their review of the research literature and current initiatives suggested "All students have access to supportive, excellent undergraduate education in science, mathematics, engineering, and technology, and all students learn these subjects by direct experience with the methods and processes of inquiry" (p. ii).

Out of the Baltimore meeting came the realization of the need to establish a National Association for Science, Technology and Society (NASTS) to promote STS education. NASTS was organized by a collection of collegiate and precollege STS enthusiasts and was officially incorporated in 1988 and headquartered within the STS Program at The Pennsylvania State University (Cutcliffe, 1996). While STS proponents found sufficient reasons to unite in a common cause, their understandings of exactly what constitutes "STS" still vary widely.

THE K–12 STS MOVEMENT IN AMERICA

The K–12 STS movement in America began in the 1960s and picked up speed in the 1970s. It spread from epicenters in private schools in New York City (via the Teachers Clearinghouse for Science Education under the direction of Irma Jarcho, John Roeder, and Nancy Van Vranken and states like Wisconsin) to become a mainstream movement in science education and technology education. To a much lesser degree, it has influenced social studies and language arts instruction. By the early 1980s STS themes were in the standard middle-level science syllabi of New York State's Regents system and in the curriculum frameworks and standards documents of other states and larger school districts.

In the United States this movement was considerably helped by the Science through Science, Technology and Society Project, which was headquartered at the Pennsylvania State University. In the 1980s it was funded by the National Science Foundation. Conferences, Curriculum modules, resource support, and a regular newsletter were provided to teachers and school systems across the United States. Papers from the annual Technological Literacy Conferences were edited and placed within the Education Resources Information Clearinghouse (ERIC) system for interested educators across the nation (Cheek and Cheek, 1996). A subsequent NSF award to Penn State established a National STS Network that created state leadership cadres of K–12 educators in 39 states using nine regional university partners across the country. These cadres, in turn, held local workshops for their peers in school systems throughout America, drawing support for their efforts from this national resource network. In K–12 education, the movement's success can be gauged by the regular appearance of STS topics in presentations and symposia at major national and regional educational conferences annually in the United States sponsored by organizations such as the National Science Teachers Association, Association for Supervision and Curriculum Development, Association for the Education of Teachers in Science, National Council for the Social Studies, National Council of Teachers of English, and the International Technology Education Association.

A deeper research question centers on the degree of STS implementation within the K–12 classrooms. The only national survey to consider this question took place in 1993. Kumar and Berlin (1996) surveyed all 50 state science supervisors regarding their perceptions of STS emphases and implementation within their respective states. They found only 17 states either required or recommended STS education as part of their science curricula. However, only three states had no STS education or what the researchers defined as "STS-Surrogate implementation." These findings are limited solely to science as a content area and rely on the perspectives of only one state education department official per state. However they do clearly signal the pervasive impact that concerns about STS education have created in K–12 American educational systems.

The K–12 STS movement recognizes the existence of technology, but it also emphasizes science content and context, while giving less time and emphasis to the technology and society aspects of the interrelationships. The isoceles triangle with "science," "technology," and "society" at each respective apex is used to symbolize STS education. In American education, that triangle has been heavily slanted toward "science."

STS in precollege education has enjoyed perhaps even more visibility and impact outside of the United States. Major curriculum reform movements in New Zealand, Australia, Canada, Great Britain, and the Netherlands, to name a few, have featured STS as prominent themes in science and technology education (Aikenhead and Solomon, 1994; Calhoun, Panwar, and Shrum, 1996; Yager, 1996). The International Network for Information in Science and Technology Education of UNESCO, UNESCO's Project 2000+, World Council on Technology Education (WCOTE), and the STS Network of the International Organization for Science and Technology Education all have STS themes and concepts in their official platforms and activities.

It is less clear what exactly constitutes STS education. Cheek (1992) surveyed a variety of influential documents and curriculum materials principally produced within the United States to ascertain commonalities among them. Out of the 30 descriptors used to discuss STS education, only six features of STS education found widespread agreement:

1. Emphasize the general interactions among science, technology and society
2. Raise levels of awareness regarding STS issues
3. Incorporate ethics and values considerations
4. Increase student understanding of the applications of technology
5. Promote decision making skills
6. Involve students in local community action.

Glen Aikenhead and Joan Solomon, influential figures in the international STS movement, have attempted to clearly define elements for STS education. (See for example, Aikenhead, 1993; Aikenhead and Solomon, 1994; Solomon, 1993.) How many United States participants would subscribe to their taxonomy remains unclear.

A national study of STS themes in state science curriculum frameworks in the United States by Kumar and Berlin (1998) suggests at least some points on which United States advocates of STS education within the science curriculum would agree with Aikenhead and Solomon's taxonomy. A total of 25 state science frameworks were examined, the majority dating to the late 1980s or early 1990s. The percentages of frameworks which addressed each of the 15 descriptors that Kumar and Berlin used to categorize STS categories within the content standards ranged from a low of 8 percent (abilities to distinguish between natural objects and objects made by humans) to a high of 88 percent (science and technology within society).

A consolation for K–12 participants is that the STS movement within American colleges and universities is also very diverse. Differing approaches and beliefs resident in college-wide STS programs target undergraduates and graduate students through specialized units that offer degrees up through the doctorate for STS studies (e.g., MIT, Rensselaer Polytechnic Institute, Penn State, Stanford).

Past Efforts at Promoting Technological Literacy

Technology, as a distinct education focus of K–12 curriculum and instruction in American schools, has had a long and checkered history. Textbooks from the 19th century contained considerable portions devoted to man's [sic] achievements in the arena of technology, often within the guise of "applied science" courses (DeBoer, 1991; Montgomery, 1994). Technological artifacts, complete with diagrams, graced many textbooks and schools engaged students in observing demonstrations and constructing simple machines and technological devices. By the late 1940s and early 1950s, public education largely turned its back on technology, and school subjects, such as science and social studies, began to be dominated by a conceptually-driven, content rich focus on disciplines as known by skilled practitioners (Welch, 1979). This was the era of the so-called "alphabet soup" science curricula with a heavy emphasis on science as known by, although not necessarily as practiced by, working scientists. Social and cultural dimensions of science were downplayed with a few notable exceptions, such as Project Physics, and technology was presented as secondary to science.

Two notable exceptions to these general trends emerged in the 1960s and 1970s. The Jackson's Mill Industrial Arts Curriculum, developed by a team of faculty centered at The Ohio State University, moved industrial arts with their emphasis on materials, construction, manufacturing, and design. Sociocultural contexts and the interrelatedness of technological systems also received attention. The history of technology per se was not in this innovative curriculum for secondary schools. Key elements of the Jackson's Mill curriculum remain embodied in materials developed by contemporary organizations in technology–vocational education such as the International Technology Education Association and from publishers such as Goodheart-Willcox, Delmar, and South-Western Publishing.

The second notable exception was *Man-Made World* a textbook by E. Joseph Piel and colleagues at the State University of New York at Stony Brook, published by McGraw-Hill (Engineering Concepts Curriculum Project, 1971). This was the first serious 20th century venture to blend science and technology in a serious manner for secondary education. It was adopted by some schools on the forefront of educational innovation, but largely ignored by most school systems in the United States. The book and its accompanying laboratory materials were influential, however, in the burgeoning STS movement in K–12 schools. It provided some of the first extended treatments of technology as a field of study for secondary schools. *Man-Made World* moved technology as a venue for student learning out of the industrial arts–vocational education wing of the high school and into the science wing—a notable achievement in itself.

STS AND K–12 SCIENCE EDUCATION

STS in K–12 American science education has taken two somewhat different paths. Some of the most vocal early advocates of STS education for the K–12 science classroom, e.g., Yager (1996; 1993) and Bybee, Carlson, and McCormack (1984), have emphasized STS as a way to teach science. In this approach, STS can stimulate student interest in science through the use of local and community STS issues, which lead students into an indepth investigation of scientific ways of understanding the world as they attempt to solve, or take informed positions on, these issues. This particular view of STS education in science education is reflected in the official position statement on STS education of the National Science Teachers Association (Yager, 1993). Even a cursory reading of contributions from this school of thought reveals a paucity of attention to the substance of technology or the substance of society (as reflected by rigorous social studies content).

A second path, which has yet to gain many adherents in the United States, has been to strongly couple technology and science as two distinct but interrelated ways of knowing and doing. The goal is to involve students in activities that demand both technological adaptation and innovation and the explicit use of scientific concepts and principles. This path was advocated by STS proponents such as Liao (1994), Roy (1990), Cheek (1992), Kumar (1998), Hurd (1997). Hurd coined the term "technoscience" to both reflect the realities of science and technology in the modern world and the need to better balance education about technology with traditional sciences in K–12 instruction and curriculum. One example of an approach in American curriculum which tries to reflect this more balanced treatment is the *Chemistry in the Community* (CHEMCOM) Project of the American Chemical Society which is now in its second edition by Kendall/Hunt Publishing. However, CHEMCOM continues to be plagued by image problems among the nation's high school chemistry teachers and thus has captured only a small part of the high school chemistry textbook market (Black and

Atkin, 1996). This path is the predominant path taken by STS in primary and secondary schools throughout most of the world, including Canada (Calhoun, Panwar, and Shrum, 1996; Weeks, 1997). Even in this arena, however, there is considerable room for improvement in the equal treatment of technology and society within the STS triangle as only cursory attention is paid to the history of technology. An indepth study of the *sociocultural* contexts of technological development is absent from precollege STS materials.

Today in American education, there is an ascendancy of standards-based approaches to curriculum, instruction, assessment, and professional development. Alignment of these four key elements of K–12 education is not only anticipated, but also required in many states, and actively promoted across a wide range of educational reform movements. There remain important tensions among national, state, and local control of the school curriculum. Many times implementation of standards within the nation's classrooms often bears only a fleeting resemblance to the new realities envisioned by the creators of standards in various content areas (Black and Atkin, 1996).

The American Association for the Advancement of Science (AAAS) launched Project 2061 in the mid 1980s with a series of blue-ribbon panels. The project's name came from the fact that a student in today's elementary school will be alive to see the return of Halley's Comet in the year 2061. Panels were charged to produce white papers on what an American high school graduate should know, value, and be able to do across a wide spectrum of human endeavor that is represented within the membership of AAAS. This includes not only the traditional science disciplines but also engineering and allied fields, social sciences, history and philosophy, education, and the arts. *Science for All Americans*, published in 1989 by AAAS, and republished with minor revisions a year later by Oxford University Press (Rutherford and Ahlgren, 1990), summarized the key findings of the blue-ribbon panels into a succinct narrative portrait of what a student should know, value, and be able to do to be considered "scientifically literate."

The history of technology and the nature of technology are treated within *Science for All Americans* (Rutherford and Ahlgren, 1990). These areas also receive considerable attention in the companion, *Benchmarks for Science Literacy* (American Association for the Advancement of Science, 1997). The Benchmarks, as they are known colloquially, contain numerous curriculum standards (benchmarks) at grades K–2, 3–5, 6–8, 9–12, which address technology in its varied dimensions, in a substantive and substantial manner. Twelve chapters in the Benchmarks document consider these topics:

1. Nature of science
2. Nature of mathematics
3. Nature of technology
4. Physical setting
5. Living environment
6. Human organism
7. Human society
8. Designed world
9. Mathematical world
10. Historical perspectives
11. Common themes (systems, models, constancy and change, scale)
12. Habits of mind

Technology was defined very broadly with the Benchmarks (AAAS, 1997) with the commentary noting:

> Technology is an overworked term. It once meant knowing how to do things— the practical arts or the study of the practical arts. But it has also come to mean innovations such as pencils, television, aspirin, microscopes, etc., that people use for specific purposes and refers to human activities such as agriculture or manufacturing and even to processes such as animal breeding or voting or war that change certain aspects of the world. Further, technology sometimes refers to the industrial and military institutions dedicated to producing and using inventions and know-how. In any of these senses, technology has economic, social, ethical, and aesthetic ramifications that depend on where it is used and on people's attitudes toward its use. (p. 43)

Chapter 3 of the Benchmarks focuses on "The Nature of Technology" distributed among three major areas: technology and science, design and systems, and issues in technology. A total of 47 benchmarks spanning four different grades (K–2, 3–5, 6–8, 9–12) provide a focus for curriculum and instruction about the nature of technology within schools. A sample benchmark for students in grades 3–5 is: "Technology enables scientists and others to observe things that are too small or too far away to be seen without them and to study the motion of objects that are moving very rapidly or are hardly moving at all (AAAS, 1997, p. 45). There are 84 benchmarks, spanning four different grades, that deal with "The Designed World" (Chapter 8) in the Benchmarks. This chapter is organized into six sections: agriculture, materials and manufacturing, energy sources and use, communication, information processing, and health technology. A series of other technology-related benchmarks are scattered across the various other chapters within the document.

The Benchmarks are probably the most extensive national attempt to create a set of curriculum content standards that cut across numerous fields of human endeavor. They provide wonderful challenges for school districts and schools to creatively weave different traditional subjects within schools in interesting and interdisciplinary ways. Unfortunately, the project has such a strong connection with "science education," due to its sponsor, that it has made little headway in school subjects outside of science education. There is little evidence that several other national curriculum standards documents in geography, history, civics, language arts, etc., created subsequent to the Benchmarks document even took account of its pioneering work in these varied school curriculum areas. A second problem has been the lack of clear agreements of respective responsibilities between the Project 2061 national staff and the three school districts with whom it has intensely worked on creating curriculum materials linked explicitly to the Benchmarks (Black and Atkin, 1996). Thus, there remains a huge gulf between the standards advocated within the document and extant curriculum materials in science or other curriculum areas.

National science standards, emanating from the National Research Council (1996) and produced by a consortium, which included AAAS and the National Science Teachers Association, were issued in 1996. Eight categories of content standards were created:

1. Unifying concepts and processes in science
2. Science as inquiry
3. Physical science
4. Life science
5. Earth and space science
6. Science and technology
7. Science in personal and social perspectives
8. History and nature of science

Considerable debate ensued over how "technology" should be treated within the standards as they were being created. Some favored simply featuring educational technology (computer hardware and software) in the science standards; others argued for a comprehensive treatment of technology similar to that seen within Project 2061. The compromise was to downplay the importance of technology as a major field of human endeavor, though it has more working practitioners by far than practicing scientists. Technology is mentioned within the document almost exclusively in connection with science and in relation to contemporary societal issues and concerns. Content Standard E (Science and Technology) within the K–4 standards states:

As a result of activities in grades K–4, all students should develop

- Abilities of technological design
- Understanding about science and technology
- Abilities to distinguish between natural objects and objects made by humans

A further standard (F: Science in Personal and Social Perspectives) includes a substatement that students should understand "science and technology in local challenges." The first two substatements of Standard E for K–4 are repeated for Standard E in grades 5–8. Standard F's substatement on STS reads "science and technology in society." Standard E for grades 9–12 repeats verbatim Standard E for grades 5–8, and the Standard F substatement now reads "science and technology in local, national, and global challenges." Many STS proponents are dissatisfied with how STS is generally treated within this important national standards document (Koch, 1996).

It is clear that technology is receiving greater attention in K–12 science education than it has since the early 1950s (DeBoer, 1991). Yet much of this attention to technology focused only on superficial aspects. Only two out of 25 state science frameworks examined by Kumar and Berlin (1998) had content standards that focused on abilities of technological design. The history of science and historical perspectives was similarly poorly represented within these framework documents. Recent framework documents from states such as Rhode Island (Rhode Island Department of Elementary and Secondary Education, 1995) and Massachusetts (Massachusetts Department of Education, 1997), however, provide more attention to technology than do many of the earlier state science frameworks. Rhode Island's emphasis on technology was a byproduct of its adoption in full of the Project 2061 *Benchmarks for Science Literacy*. Massachusett's attention to technology was due to the sustained advocacy and involvement of the Technology Education Association of Massachusetts, Inc. and its national organization, the International Technology Education Association.

The National Science Foundation (NSF) has had an abiding interest in the history of science and history of technology as fields of scholarly research (Jasanoff, Markle, Peterson, and Pinch, 1995). The NSF has funded projects that use the history of science as a vehicle to increase student understanding of the processes of science. More recently, the foundation has turned increasing attention to technology education, thanks to the advocacy efforts of Gerhard Salinger, a physicist from Rensselaer Polytechnic Institute in Troy, NY. Through its Advanced Technological Education and Instructional Materials Development programs, NSF has been a primary funder of the

Technology for All Americans Project (see below) and a series of projects designed to target either Technology Education (TE), Science and Technology Education (STE), or Mathematics, Science and Technology Education (MSTE). The NSF undertook a comprehensive study of middle school science materials they funded. Their review was released in February 1997. They found that the materials reviewed lacked enough focus on the history and nature of science (National Science Foundation, 1997). This same comment certainly applies to technology, which was not even considered by the review panel as a criterion for evaluation!

STS in K–12 Social Studies

The STS movement in social studies as taught in American schools can generally be characterized as much talk but very little action. There have been several attempts to promote STS education in social studies through the efforts of a now dissolved Science and Society Committee within the National Council for the Social Studies (NCSS). Committee members for the past decade or more have organized and facilitated presentations and symposia at annual national and state meetings of NCSS (Splittgerber, 1996). *Social Education*, the flagship publication of NCSS, carried substantial articles from time to time about STS education in social studies.

The explicit mention of STS as one of the ten essential themes in social studies in the *Curriculum Standards for the Social Studies* promulgated by NCSS (1994) was a clarion call to focus on this important arena of human societies within social studies programs. STS takes its place alongside the following key social studies themes:

1. Culture
2. Time, continuity and change
3. People, places, and environment
4. Individual development and identity
5. Individuals, groups, and institutions
6. Power, authority, and governance
7. Production, distribution and consumption
8. Global connections
9. Civic ideals and practice

The strand is explained as follows

Technology is as old as the first crude tool invented by prehistoric humans, but today's technology forms the basis for some of our most difficult social choices. Modern life as we know it would be impossible without technology and the science that supports it. But technology brings with it many questions: Is new technology always better than that which it will replace? What can we learn from the past about how new technologies result in broader social change, some of which is unanticipated? How can we cope with the ever-increasing pace of change, perhaps even with the feeling that technology has gotten out of control? How can we manage technology so that the greatest number of people benefit from it? How can we preserve our fundamental values and beliefs in a world that is rapidly becoming one technology-linked village? This theme appears in units or courses dealing with history, geography, economics, and civics and government. It draws upon several scholarly fields from the natural and physical sciences, social sciences, and the humanities for specific examples of issues and the knowledge base for considering responses to the societal issues related to science and technology.

Young children can learn how technologies form systems and how their daily lives are intertwined with a host of technologies. They can study how basic technologies such as ships, automobiles, and airplanes have evolved and how we have employed technology such as air conditioning, dams, and irrigation to modify our physical environment. From history (their own and others), they can construct examples of how technologies such as the wheel, the stirrup, and the transistor radio altered the course of history. By the middle grades, students can begin to explore the complex relationships among technology, human values, and behavior. They will find that science and technology bring changes that surprise us and even challenge our beliefs, as is the case of discoveries and their applications related to our universe, the genetic basis of life, atomic physics, and others. As they move from the middle grades to high school, students will need to think more deeply about how we can manage technology so that we control it rather than the other way around. There should be opportunities to confront such issues as the consequences of using robots to produce goods, the protection of privacy in the age of computers and electronic surveillance, and the opportunities and challenges of genetic engineering, test-tube life, and medical technology with all their implications for longevity and quality of life and religious beliefs. (NCSS, 1994, p. 28)

Sixteen different performance expectations related to student knowledge of STS were delineated to accompany the general description of the theme: five for the early grades, five for the middle grades, and six for high school. Science and technology are inextricably coupled in the phraseology within these expectations, with the exception of three that treat only one or the other arena. Thus we see in this document an important emphasis on "reclaiming science [and technology] for social knowledge" as Fleury (1997) characterizes it, but considerably less attention to defining and considering technology as distinct from science.

The Social Studies Development Center and ERIC Clearinghouse for Social Studies and Social Education at Indiana University under John Patrick and associates, have worked hard to solicit STS manuscripts and curriculum materials in social studies and place them within the ERIC system. A series of STS education products from the Social Sciences

Education Consortium in Boulder, CO (SSEC) has provided teachers with lessons that could be adopted or adapted. For additional background materials to increase teacher knowledge and awareness of STS issues in the social studies, see Singleton (1995) and Biological Sciences Curriculum Study, Social Sciences Education Consortium (1992).

NSF funded the Lemelson Center of the Smithsonian Institution and the Society for the History of Technology to develop jointly materials on Discovering Science and Technology through History (The website for the project is http://www.si.edu/lemelson).

This project, directed by historian of technology Susan Smulyan of Brown University, is a curriculum for secondary classes, principally in social studies. Its central theme of textiles, dyeing, and industrialization was chosen for three principal reasons:

1. Industrialization is central in United States history and the textile industry was a key component of United States industrialization.
2. Clothing appeals to teenagers as a subject, while providing ample opportunities to bridge history, social studies, technology, and science.
3. Such a topic might especially interest young women and minority students.

The project has eight curriculum units, which include hands-on activities. These were released in a field-test version in 1998. In one unit, students learn about economics and industrialization by focusing on the Civil War era. Students lay out patterns for uniforms and calculate the cost of mass production. There are primary documents to interpret, graphs and charts to create and understand, and ideas for games, debates, and class discussions. Scholarly articles, historical essays, a bibliography, and a videotape on textile machinery complete the package. It remains to be seen what type of following these materials will attract among the nation's high school social studies teachers. Despite the worth of these particular materials, anyone sensitive to the history of technology in its sociocultural contexts would be disappointed with the cursory treatment technology gets in these materials promulgated for the social studies classroom.

STS IN TECHNOLOGY EDUCATION

Technology education is the final major arena within K–12 American education where attention to technology has thrived. The modern technology education movement, as represented by the International Technology

Education Association (ITEA) headquartered in Reston, VA, emerged from the American Vocational Education Association. ITEA has commissioned, published, and distributed a variety of publications focused explicitly on technology education for K–12 schools, including their flagship publication, *The Technology Teacher*. Research findings related to technology education have appeared in the *Journal of Technology Education*, a refereed journal published in conjunction with the Council on Technology Teacher Education.

Technology education continues to have a marginalized existence in the American secondary school curriculum. This is due to its long and historic association with vocational education and a divide that opened up between academic and career education in the early 20 century, much to the chagrin of John Dewey and the progressives (Wirth, 1972). Today's technology education in high schools usually resides in the career–vocational wing of the building. Courses are often taught by traditional "shop" teachers who now have proclaimed themselves as "technology educators." Yet they lack sufficient professional development and course preparation to make the needed transitions in subject matter knowledge and skills.

The place of technology within the American school curriculum is at variance with its position in the curricula of other industrialized nations. The Netherlands, New Zealand, Australia, Canada, and the United Kingdom are representative examples where technology has a much larger niche within primary and secondary education than it possesses within America (DeVries, Cross, and Grant, 1993; Gordon, Hacker, and DeVries, 1993; Eggleston, 1996). In England, for example, design and technology is a key component within the national curriculum, separate and distinct from science education (Eggleston, 1996; Kimbell, Stables, and Green, 1996). David Layton (1993) professor emeritus of Science Education at Leeds University in England has been instrumental in the ascendancy of technology within the national curriculum. He argues that technology education poses significant challenges to science education to overhaul its approach in the primary and secondary curriculum.

A host of curriculum materials for technology education in America is produced, principally by small publishers who have a market niche in this arena. Only a flavor of these materials can be provided here. Traditionally, American technology education features four or five main topics in its textbooks: manufacturing, communications, transportation, construction, biotechnology, and energy (Gallo, Somon, and Swernofsky, 1993 Gradwell, Welch, and Martin, 1996; Lister, 1987; Pierce and Karwartka, 1993; Williams, Badrhkan, and Daggett, 1985; Wright, 1996). Student activities include building simple or increasingly sophisticated technological devices. Students sometimes create timelines or brief summaries of the history of a particular invention, generally centered on a particular artifact rather than an entire technological system. There is no indepth look at the evolution of a particular technology, or even a move beyond the internalist tradition within the history of technology. Technology education still has a long way to go in America to realize the dream of STS as a bridge to solidify fairly equal attention to both science and technology (Zuga, 1991; 1996).

One positive development in this arena has been the growth and improvement in recent years of "Tech Prep" programs. The Perkins Vocational Education Act, enacted by Congress in the late 1980s, stimulated the development of programs for the last two years of high school that would directly connect with subsequent two year associate degree programs in technical areas at community colleges. Many of these courses in the early days were developed by a consortium of states through the Curriculum Occupational Research and Development Center in Texas (CORD) headed by Dan Hull (1993; Hull and Purnell, 1991). CORD materials were the first wave of curriculum materials to teach physics, chemistry, communications (Language arts), and mathematics in an applied manner targeted to students whose career goals might center on skills developed via a two year technical college degree. Recently, there has been a widening of the Tech Prep concept at both federal and state levels to embrace students from a wider variety of prospective future careers. Connections between Tech Prep and the newly ascendant School-to-Work movement funded by various federal agencies to substantially overhaul career and vocational education in the nation are still being worked out in each individual state. Unfortunately, most of the Tech Prep materials to date focus on representative technologies of importance to particular career paths. There is insufficient attention to the social and cultural contexts of the technologies in question.

This cursory survey of extant materials in technology education reveals once again a consistent picture of inattention to the sociocultural history of technology in America and modern life (Pursell, 1995; Hays, 1995). An article by McGee and Wicklein (1997) in *The Technology Teacher* recognized that technology education in K–12 schools largely ignored the history of technology. A middle school teacher, Nancy Matheny, is quoted as saying, "Teaching technology education without a historical component is kind of like teaching art without delving into art history. Art, like technology, is more than three times as old as writing and contains, in its changing styles and themes, a 15,000-year-old record of the physical and social evolution of man" (p. 18). The authors conclude, "To truly educate our students within our field, the concept of technology's history must be integral in the technology education curriculum" (McGee and Wicklein, 1997, p. 18). It remains to be seen whether curriculum developers of technology education can, or will, rise to this challenge.

Notably, Project 2061 interacted little with ITEA organizationally in the creation of *Science for All Americans* or the *Benchmarks for Science Literacy*. Project 2061's minimal interaction, along with each professional subject matter group in education trying to create its own "national" standards, forced ITEA to obtain funding from the National Science Foundation and NASA to produce a set of curriculum standards for technology education. This was appropriately titled, "Technology for All Americans." A national advisory board consists of representatives from other organizations who have produced national standards documents in science and mathematics as well as members from business and industry and federal and state government.

A historian of technology from the Henry Ford Museum chaired the committee that produced the initial guiding document for the project, *Technology for All Americans: A Rationale for Technology Education* (International Technology Education Association, 1996). The document defines "technological literacy" as "the ability to use, manage, and understand technology." Each of the verbs is then further defined as follows:

> The ability to *use* technology involves the successful operation of the key systems of the time. This includes knowing the components of existing macro-systems, or human adaptive systems, and how the systems behave.

> The ability to *manage* technology involves insuring that all technological activities are efficient and appropriate.

> *Understanding* technology involves more than facts and information, but also the ability to synthesize the information into new insights. (p. 6)

The universals of technology are defined as centering on processes, knowledge, and contexts. *Processes* include the design, development, use, control, assessment and consequences of technological systems. *Knowledge* involves the nature and evolution of technology, linkages with other arenas of human activity and knowledge, and technological concepts and principles. *Contexts* are segmented into informational systems, physical systems, and biological systems. An instrumentalist and internalist approach to technology is apparent within the document.

A decision was made within the ITEA project that all members of the writing committee for the guide document should not serve on the standards committee that would actually draft what every student in America should know, value, and be able to do in the area of technology. The project released its second draft of technology standards for K–12 education for public review, hearings, and focus groups in spring 1998. A third draft became available in fall 1998 with an expectation that a final revision will occur in light of field testing in the fall. A fourth draft was produced in the summer of 1999 and selectively distributed. A panel of experts was convened in August, 1999, under the auspices of the National Academy of Engineering within the National Research Council to review the draft. A final revision of the standards is projected to be released in the spring of 2000. The writing team for the standards is composed of teachers of technology education in K–12 schools, district and state technology education supervisors, university faculty in technology education, and a limited number of business and industry representatives. Unfortunately there is no representation on the writing team of experts in the fields of philosophy, history, anthropology, or sociology of technology.

Beyond the creation of "national" standards for technology education and curriculum materials linked to those standards lies the daunting hurdle of implementation of these materials in the K–12 school systems of this nation. Certainly the experience of curriculum developers is insightful, and it is to this perspective that we now turn.

THE PERENNIAL CHALLENGES OF EDUCATIONAL REFORM

Philip W. Jackson (1983) reported that, in the period 1954–1975, the NSF funded 53 projects to the tune of $117 million. Wayne Welch (1979) using a slightly different method of tabulation, reported a total of $130 million for course content improvement projects and another $565 million for teacher training activities during that same period. While these numbers appear large, it is important to realize that federal funding varied greatly within that time, and the relative size of these figures is small compared with the over $100 billion spent annually on K–12 education during study the period. Despite, or maybe in part because of, these moderate investments, these curricula had only limited impact in changing classroom instruction and student achievement. Welch (1979) noted that "Curriculum does not seem to have much impact on student learning no matter what curriculum variations were used. . . . we at Project Physics eventually concluded that 5 percent [variance in student achievement of old versus new curriculum] was an acceptable return on our investment since we could seldom find greater curricular impact on the students" (p. 301).

Welch was deeply involved in Project Physics efforts centered at Harvard University to teach physics with a heavy historical flavor. A series of central factors mitigated change including:

- The unwillingness of course developers (usually university faculty) to listen to teacher suggestions for revisions of the materials
- The narrow federal funding timetable that demanded completed projects within a three to four year timespan

- An attempt to produce "teacher-proof" curricula
- An overemphasis on the discipline as known by the skilled practitioner rather than the entering novice (cf. DeBoer, 1991)

In addition Welch observed:

> From the beginning, there were the known challenges of unprepared and insecure science teachers, the inherent difficulty of change, the lack of federal policy for innovations, the natural conservatism of schools, and the threat of a national curriculum. But in the second decade were added the unforeseen problems of declining enrollments at the secondary level, inflation, student unrest, a fading public image of science, environmental concerns, competing demands such as integration, the back-to-basics movement, social concerns, and school reform movement. (Welch, 1979, p. 292)

CONCLUSION

Technology is a pervasive influence in the modern world. We still have not reached a point in K–12 education, even within the STS paradigm, where technology as a subject for study, investigation, and thought has received its due place. As an NSF (1996) report reminds us: "In an increasingly technical and competitive world with information as its common currency, a society without a properly educated citizenry will be at great risk and its people denied the opportunity for a fulfilling life" (p. ii).

Education policymakers must exert fuller efforts to ensure that the divide between science and technology is bridged and that technology begins to occupy its rightful place in both the K–12 curriculum and the learning experience of students. This requires an end to the hegemony of science educators and scientists who control the technology discourse within K–12 education circles. A greater involvement of engineers and technology educators is needed to remedy the deficiencies outlined in this chapter. A second major barrier to increased attention to technology within STS education is the continuing inadequacy of teacher preparation programs in science, social studies, and the language arts which continue to produce graduates who lack substantive coursework or experience with technology beyond computing. This will require changes to the teacher certification course requirements within all 50 states and substantial upgrading of college faculty's own knowledge about technology and its multifaceted nature. Further success depends on research into the ways science and technology education can cooperate together in the interests of better educating all students about the pervasive technological world which we inhabit. Much of this research needs to take the form of action research with open-ended, exploratory questions being posed, attempted, analyzed, refined etc. in continuous iterations concerning relationships between science and technology and potential synergy arising from the use of concepts and methods within these respective fields of human endeavor. Several research journals are partially devoted to exploring these relationships: *Journal of Technology Education*, *Journal of Design and Technology*, *Research in Science and Technological Education*, and the *Journal of Science Education and Technology*. We owe it to our children and future generations to broaden and deepen attention to technology within K–12 education in the United States and around the globe.

ACKNOWLEDGMENTS

Several key ideas in this chapter were originally presented in the Symposium "New Ideas, Audiences, and Venues for Teaching the History of Technology," at the Society for the History of Technology Annual Meeting in Pasadena, CA, October 15–17, 1997. The author gratefully acknowledges the efforts of Ed Pershing (coordinator), Steve Cutcliffe (chair), fellow presenters, and participants.

REFERENCES

Aikenhead, G. (1993). *Logical Reasoning in Science and Technology*, John Wiley and Sons of Canada, Toronto.

Aikenhead, G., and Solomon, J. (eds.) (1994). *STS Education: International Perspectives on Reform*, Teachers College Press, New York.

American Association for the Advancement of Science (1997). *Benchmarks for Science Literacy*, Oxford University Press, New York.

Bauer, M. (ed.) (1997). *Resistance to New Technology. Nuclear Power, Information Technology and Biotechnology*, Cambridge University Press, New York.

Billington, D. P. (1996). *The Innovators: The Engineering Pioneers Who Made America Modern*, John Wiley and Sons, New York.

Biological Sciences Curriculum Study, Social Sciences Education Consortium (1992). *Teaching about the History and Nature of Science and Technology: A Curriculum Framework*, Boulder, CO.

Black, P. J., and Atkin, M. (eds.) (1996). *Changing the Subject: Innovations in Science, Mathematics, and Technology Education*, Routledge, New York.

Bybee, R. W., Carlson, J., and McCormack, A. J. (eds.) (1984). *NSTA Yearbook: Redesigning Science and Technology Education*, National Science Teachers Association, Washington, D.C.

Calhoun, K., Panwar, R., and Shrum, S. (eds.) (1996). *International Organization for Science and Technology Education (IOSTE) 8th Symposium Proceedings. Four volumes*, Faculty of Education, University of Alberta, Edmonton, Alberta, Canada.

Cheek, D. W. (1992). *Thinking Constructively about Science, Technology and Society Education*, State University of New York Press, Albany, NY.

Cheek, D. W., and Cheek, K. A. (eds.) (1996). *Proceedings of the Eleventh National Technological Literacy Conferences, February 8–11, 1996, Arlington, VA*. ERIC Clearinghouse for Social Sciences and Social Studies Education, Bloomington, IN.

Cutcliffe, S. H. (1996). National Association for Science, Technology, and Society. In Yager, R. E. (ed.), *Science/Technology/Society as Reform in Science Education*, State University of New York Press, Albany, NY, pp. 291–297.

DeBoer, G. E. (1991). *A History of Ideas in Science Education: Implications for Practice*, Teachers College Press, New York.

DeVries, M. J., Cross, N., and Grant, D. P. (eds.) (1993). *Design Methodology and Relationships with Science*, Springer-Verlag, New York.

Eggleston, J. (1996). *Teaching Design and Technology* (Second Edition), Open University Press, Philadelphia.

Ellul, J. (1990). *The Technological Bluff*, Wm. B. Eerdmans Publishing, Grand Rapids, MI.

Engineering Concepts Curriculum Project (1971). *The Man-Made World*, McGraw-Hill, New York, NY.

Fleury, S. C. (1997). Science in social studies: Reclaiming science for social knowledge. In Ross, E. W. (ed.), *The Social Studies Curriculum: Purposes, Problems, and Possibilities*, State University of New York Press, Albany, NY, pp. 165–182.

Gallo, D., Somon, S., and Swernofsky, N. R. (1993). *Experience Technology*, Glencoe Publishers, New York.

Gordon, A., Hacker, M., and DeVries, M. (eds.) (1993). *Advanced Educational Technology in Technology Education*, Springer-Verlag, New York.

Gradwell, J., Welch, M., and Martin, E. (1996). *Technology Shaping Our World*, Goodheart-Wilcox Company, Inc., South Holland, IL.

Green, E., Owen, J., and Pain, D. (1993). *Gendered by Design? Information Technology and Office Systems*, Taylor and Francis, Washington, D.C.

Hays, S. P. (1995). *The Response to Industrialism, 1885–1914* (Second Edition), University of Chicago Press, Chicago.

Hull, D. (1993). *Opening Minds, Opening Doors: The Rebirth of American Education*, Center for Occupational Research and Development, Waco, TX.

Hull, D., and Parnell, D. (1991). *Tech/Prep/Associate Degree: A Win/Win Experience*, Center for Occupational Research and Development, Waco, TX.

Hurd, P. D. (1997). *Inventing Science Education for the New Millennium*, Teachers College Press, New York.

International Technology Education Association (1996). *Technology for All Americans: A Rationale for Technology Education*, Reston, VA.

Jackson, P. W. (1983). The reform of science education: A cautionary tale. *Daedalus*, 112(2):143–166.

Jasanoff, S., Markle, G. E., Peterson, J. C., and Pinch, T. (eds.) (1995). *Handbook of Science and Technology Studies*, Sage Publications, Thousand Oaks, CA.

Kimbell, R., Stables, K., and Green, R. (1996). *Understanding Practice in Design and Technology*, Open University Press, Philadelphia, PA.

Kline, S. J. (1985). What is technology? *Bulletin of Science, Technology and Society* 5(3):215–218.

Koch, J. (1996). National science education standards: A turkey, a valentine, or a lemon? In Yager, R. E. (ed.), *Science/Technology/Society as Reform in Science Education*, State University of New York Press, Albany, NY, pp. 306–315.

Kumar, D. D. (1998). Chemical education in an era of information technology. *The Chemist* 75(1):3–4.

Kumar, D. D., and Berlin, D. F. (1998). A study of STS themes in state science curriculum frameworks in the United States. *Journal of Science Education and Technology* 7(2):191–197.

Kumar, D. D., and Berlin, D. F. (1996). A study of STS curriculum implementation in the United States. *Science Educator* 4(1):12–19.

Latour, B. (1996). *ARAMIS or the Love of Technology*, Harvard University Press, Cambridge, MA.

Layton, D. (1993). *Technology's Challenge to Science Education*, Open University Press, Philadelphia, PA.

Liao, T. (1994). Principles of engineering course: School and industrial collaboration. In Blandow, D. and Dyrenfurth, M. J. (eds.) *Technology Education in School and Industry: Emerging Didactics for Human Resource Development*, Springer-Verlag, New York, pp. 197–207.

Listar, G. (1987). *Technology Activity Guide 1*, Delmar Publishers, Albany, NY.

MacKenzie, D. (1996). *Knowing Machines: Essays on Technical Change*, MIT Press, Cambridge, MA.

Marcus, A. I., and Segal, H. P. (1989). *Technology in America: A Brief History*, Harcourt, Brace, Jovanovich, San Diego, CA.

Massachusetts Department of Education (1997). *Science and Technology Curriculum Framework. Owning the Questions through Science and Technology*, Malden, MA.

McAdams, R. (1996). *Path of Fire: An Anthropologist's Inquiry into Western Technology*, Princeton University Press, Princeton, NJ.

McGee, L., and Wicklein, R. C. (1997). Technology education in perspective: Clearer visions necessary. *The Technology Teacher* October:17–20.

McGinn, R. E. (1991). *Science, Technology and Society*, Prentice Hall, Englewood Cliffs, NJ.

Melzer, A. M., Weinberger, J., and Zinman, M. R. (1993). *Technology in the Western Political Tradition*, Cornell University Press, Ithaca, NY.

Mitcham, C. (1994). *Thinking through Technology: The Path Between Engineering and Philosophy*, University of Chicago Press, Chicago.

Montgomery, S. L. (1994). *Minds for the Making: The Role of Science in American Education, 1750–1990*, The Guilford Press, New York.

Morgall, J. M. (1993). *Technology Assessment: A Feminist Perspective*, Temple University Press, Philadelphia, PA.

National Council for the Social Studies (1994). *Curriculum Standards for Social Studies: Expectations of Excellence*, Washington, D.C.

National Research Council (1996). *National Science Education Standards*, Washington, D.C.

National Science Foundation (1996). *Shaping the Future: New Expectations for Undergraduate Education in Science, Mathematics, Engineering, and Technology*, NSF96-139, Arlington, VA, p. ii.

National Science Foundation (1997). *Review of Instructional Materials for Middle School Science*, Directorate for Education and Human Resources, Arlington, VA.

National Science Teachers Association (1982). *Science–Technology–Society: Science Education for the 1980s*, Washington, D.C.

Nye, D. E. (1994). *American Technological Sublime*, MIT Press, Cambridge, MA.

Pierce, A., and Karwatka, D. (1993). *Introduction to Technology*, West Publishing Company, St. Paul, MN.

Pound, N. J. G. (1989). *Hearth and Home: A History of Material Culture*, University of Indiana Press, Bloomington, IN.

Pursell, C. (1995). *The Machine in America: A Social History of Technology*, The Johns Hopkins University Press, Baltimore, MD.

Rhode Island Department of Elementary and Secondary Education (1995). *Science Literacy for All Students: The K–12 Science Framework*, Providence, RI.

Rothenberg, D. (1993). *Hand's End: Technology and the Limits of Nature*, University of California Press, Berkeley, CA.

Roy, R. (1990). The relationship of technology to science and teaching of technology. *Journal of Technology Education* 1(2):5–19.

Rutherford, J., and Ahlgren, A. (1990). *Science for All Americans*, Oxford University Press, New York.

Sarewitz, D. (1996). *Frontiers of Illusion: Science, Technology and the Politics of Progress*, Temple University Press, Philadelphia, PA.

Sclove, R. E. (1995). *Democracy and Technology*, Guilford Press, New York.

Segal, H. P. (1994). *Future Imperfect: The Mixed Blessings of Technology in America*, University of Massachusetts Press, Amherst, MA.

Simpson, L. C. (1995). *Technology, Time, and the Conversations of Modernity*, Routledge, New York.

Singleton, L. R. (1995). *Science/Technology/Society: Activities and Resources for Secondary Science and Social Studies*, Social Sciences Education Consortium, Boulder, CO.

Solomon, J. (1993). *Teaching Science, Technology and Society*, Open University Press, Philadelphia.

Splittgerber, F. (1996). Science–technology–society themes in social studies. *Theory into Practice* 30(4):242–250.

Stevens, E. W. Jr. (1995). *The Grammar of the Machine: Technical Literacy and Early Industrial Expansion in the United States*, Yale University Press, New Haven, CT.

Volti, R. (1995). *Society and Technological Change* (Second Edition), St. Martin's Press, New York.

Wajcman, J. (1991). *Feminism Confronts Technology*, The Pennsylvania State University Press, University Park, PA.

Waks, L. J. (ed.) (1987). Technological Literacy: Proceedings of the Second National Conference, Washington, D.C., February 6–8, 1987. *Special Issue of the Bulletin of Science, Technology and Society* 7(1, 2):1–366.

Webster, A. (1991). *Science, Technology and Society*, Rutgers University Press, New Brunswick, NJ.

Weeks, R. C. (1997). *The Child's World of Science and Technology: A Book for Teachers*, Prentice Hall Allyn and Bacon, Scarborough, Ontario.

Welch, W. W. (1979). Twenty years of science curriculum development: A look back. In Berliner, D. C. (ed.), *Review of Research in Education*, Volume 7, American Educational Research Association, Washington, D.C.

Wenk, E. Jr. (1995). *Making Waves: Engineering, Politics, and the Social Management of Technology*, University of Illinois Press, Urbana, IL.

Westrum, R. (1991). *Technologies and Society: The Shaping of People and Things*, Wadsworth Publishing, Belmont, CA.

Williams, C. F., Badrhkan, K. S., and Daggett, W. R. (1985). *Technology for Tomorrow*, University of Chicago Press, Chicago.

Winner, L. (1986). *The Whale and the Reactor: A Search for Limits in an Age of High Technology*, University of Chicago Press, Chicago, IL.

Wirth, A. G. (1972). *Education in the Technological Society: The Vocational-Liberal Studies Controversy in the Early Twentieth Century*, Intext Educational Publishers, Scranton, PA.

Wright, T. R. (1996). *Technology Systems*, Goodheart-Wilcox Company, Inc., South Holland, IL.

Yager, R. (ed.) (1993). *What Research Says to the Science Teacher. Volume Seven: The Science, Technology, Society Movement*, National Science Teachers Association, Washington, D.C.

Yager, R. E. (ed.) (1996). *Science/Technology/Society as Reform in Science Education*, State University of New York Press, Albany, NY.

Zuga, K. F. (1991). The technology education experience and what it can contribute to STS. *Theory into Practice* 30(4):260–266.

Zuga, K. F. (1996). STS promotes the rejoining of technology and science. In Yager, R. E. (ed.), *Science/Technology/Society as Reform in Science Education*, State University of New York Press, Albany, NY, pp. 227–240.

CHAPTER 8

Student Understanding of Global Warming
Implications for STS Education beyond 2000[1]

James A. Rye and Peter A. Rubba

Many experts believe humans are imperiling the ecology of the earth by enhancing the natural greenhouse effect, which may result in global warming. Others suggest, however, that we do not yet fully understand all the factors operating in the earth's system and their complex interactions, so it is possible that the warming observed during the past century may be due to natural variation. Whether or not there is a discernible human influence

---

[1] This chapter is based upon work supported by the National Science Foundation under grant no. TEP-9150232. Any opinions, findings, and conclusions or recommendations expressed in this material are those of the authors and do not necessarily reflect the views of the National Science Foundation.

James A. Rye, Department of Educational Theory and Practice, West Virginia University, Morgantown, WV 26506-6122.   Peter A. Rubba, Department of Curriculum and Instruction, The Pennsylvania State University, University Park, PA 16802.

Science, Technology, and Society: A Sourcebook on Research and Practice, edited by Kumar and Chubin, Kluwer Academic / Plenum Publishers, New York, 2000.

on global climate, global warming is an STS issue that will continue to be debated far into the next century. The potential significance of global warming makes it the STS issue that will shape STS education as STS enters the new millennium.

Separating the effects of natural variation in climate from the possible impact of human influences is an extremely complex problem. Still, with each passing year the evidence mounts that the enhanced greenhouse effect is a serious threat to our biosphere and to the human economies responsible for the consequent global warming. Nations around the world have industrialized power provided by the burning of fossil fuels. In the process, carbon dioxide and water vapor (the major byproducts of fossil fuel burning), and other gases related to industrial activity, such as methane, nitrous oxide, and chlorofluorocarbons (CFCs), are vented into the atmosphere where they can trap heat and enhance the earth's natural greenhouse effect.

The sharp increases in these "greenhouse" gases, particularly carbon dioxide, over the past 150 years have corresponded to increases in global temperatures. The year 1997 was the hottest year on record, with the nine hottest years on record occurring within the last 11, according to data collected by the National Oceanic and Atmospheric Administration (NOAA) (Warrick, 1998). Now, global and regional changes which may result from an enhanced greenhouse effect are being identified. These include, for example, retreating glaciers and melting ice caps, increases in the number and severity of storms, diminished growth in coral reefs, marked seasonal changes in the northern latitudes (spring occurring earlier and fall later), and changes in climate-related disease distribution—unprecedented malaria, hantavirus and cholera outbreaks in climates where these diseases usually do not occur (O'Meara, 1997). More dramatic and potentially catastrophic signs of climatic change, such as a loss in biodiversity, the flooding of coastal areas, regional changes in soil moisture content and food production (Ennis and Marcus, 1994; Mackenzie and Mackenzie, 1995; Schneider, 1989), are less easily separated from natural variability (Houghton, Callendar, and Varney, 1992). New data continue to strengthen the convictions of scientists who believe that increased greenhouse gas levels in the atmosphere are contributing to global warming, and to sway other scientists. In an Associated Press interview, Elbert Friday, a meteorologist and a research chief at NOAA said, "I wouldn't have been willing to say this two years ago. I believe we are seeing evidence of global warming at least some of which is attributable to human activities" (Schmid, 1998, p. A8).

Accordingly, global climate change has received considerable international attention from scientists and policymakers. Profound examples of such attention include the establishment of the Intergovernmental Panel on

Climate Change (IPCC) in 1988 by the World Meteorological Organization and the United Nations Environment Program, the 1992 United Nations Framework Convention on Climate Change, and the December 1997 Kyoto Climate Summit. There has been a corresponding interest in educating youth through college students about global climate change (e.g., Activities for the Changing Earth System, 1993; Mackenzie, 1998; Roleff, 1997; Science for Understanding Tomorrow's World: Global Change, 1994; see also University Corporation for Atmospheric Research on-line at http://www.ucar.edu). Moreover, this interest is likely to gain momentum in light of the IPCC forecasts (Houghton, Filho, Callander, Harris, Kattenberg, and Maskell, 1996) for future climate warming and the projected impact of such on the environment and society, as described subsequently.

    The IPCC predicts an increase in global mean temperature of about 2°C by the year 2100; the range of predicted increases is 1–3.5°C, which reflects low to high greenhouse gas emission and climate sensitivity scenarios (Kattenberg, Giorgi, Grassl, Meehl, Mitchell, Stouffer, Tokioka, Weaver, and Wigley, 1996). The "best estimate" of corresponding sea level rise by the year 2100—the majority due to thermal expansion of the oceans as opposed to the melting of glaciers and ice caps—is about 50 cm (Warrick, Provost, Meier, Oerlemans, and Woodworth, 1996). Considerable scientific uncertainty surrounds these estimates of future rise in sea level: The range is 20 cm to 86 cm for the previously cited "best estimate." Indeed, scientific uncertainties are inherent in all predictions of climate change and its impact. Nevertheless, a sea level rise of about 50 cm will "lead to inundation of low-lying areas around the world" (Mackenzie, 1998, p. 396) and double the number of people who currently are at risk for flooding (IPCC, 1995, Metzger, 1996); rises of 100 cm, a value close to the highest IPCC Working Group I estimate, will have a profound impact on the lives of the 20 percent of the world's inhabitants who populate these areas. Such an impact would include reducing land surface area, tourism and fresh water supplies, the latter already at a premium for some islands and low-lying coastal areas (Mackenzie, 1998, IPCC, 1995).

    Kotlyakov (1996) describes the varying impact by geographic region that is expected to accompany global warming, e.g., maximum warming will occur at high latitudes and result in relative losses of permafrost and the Greenland ice sheet whereas the midlatitudes of the northern hemisphere may reap benefits from increased agricultural production. The IPCC (1995) also acknowledges that climate warming likely will benefit some geographic regions; however, they summarize the social costs of anthropogenically induced warming by stating that such beneficial impacts are "dominated by the damage costs" (p. 50). For example, human health generally will be adversely impacted by these results of climate warming: increased morbidity and mortality due to heat stress, vector-borne infectious diseases (malaria, viral encephalitis, etc.), and nonvector-borne diseases such as cholera and giardiasis. Kotlyakov (1996) predicts climate warming will result in an increased frequency of extreme phenomena (such as drought) and believes "society is about to enter a period of environmental and economic adjustment to rapidly changing conditions" (p. 522).

The observed and predicted impacts of increased greenhouse gas emissions, such as those described above, make global climate change issues particularly appropriate themes for STS education. Such themes are timely and of interest to students; they are of global significance yet have local implications, so the possibility of taking action exists. Additionally, they are linked to science concepts typically studied in middle and secondary school science courses, and the diversity of these concepts—such as photosynthesis, electromagnetic spectrum, and the earth surface reservoirs of the atmosphere, hydrosphere, lithosphere, and biosphere—provides an excellent opportunity to integrate instruction on the life, physical, earth, and space sciences. STS themes on global climate change issues also provide a rich context for exploring "the conflict" that exists between the human-made "technosphere" (p. 6) and the natural earth surface reservoirs (Mackenzie, 1998).

Yet another factor that makes climate change issues, specifically global warming, especially suited for STS themes is that "the reality question [of global warming] is still under some debate" (Kerr, 1997, p. 1917). Further, Mackenzie (1998) reports that "there is substantial disagreement in the scientific community concerning the environmental problems of an enhanced greenhouse effect..." (p. 416). Whereas such disagreement among scientists results in a quandary and debate among economists, policymakers, journalists—society in general—it also provides an opportunity for students to consider and think critically about opposing viewpoints (Roleff, 1997). Metzger (1996) contends that such controversy taken together with the potential dramatic effects of global climate change uniquely positions this topic for "exploring the dynamic nature of science and its importance to society" (p. 327).

Over the past five years, we have had the privilege of working with colleagues, middle–junior high school science teachers and their students to develop STS curricula on the science, technology, and societal issues related to certain critical aspects of global climate change, and to study both students and teachers during the development and implementation of the curricula. This work focused mainly on the concepts and STS issues of global warming and ozone layer depletion, and to a lesser degree on ground level ozone pollution. In this chapter, we draw upon the literature and our published work, together and with others as cited herein, to discuss students' pre- and post-instructional alternative concepts of global warming, and the implications the findings have for STS education.

A Leadership Institute in Science–Technology–Society (STS) Education, funded by the National Science Foundation, served as the mechanism for our work with two dozen mainly middle–junior high school science teachers from rural central Pennsylvania and northern West Virginia and their students. The institute's program was based on a research foundation that connects it to the long acknowledged scientific literacy goal of K-12 science education.

SCIENTIFIC LITERACY AS A FOUNDATION FOR STS EDUCATION

The term "scientific literacy" was coined after World War II. Although our concepts of scientific literacy have changed over time, our preparation of citizens to deal with science and technology as these enterprises touch their lives has been a generally acknowledged goal for a school science education since Benjamin Franklin and Thomas Jefferson's advocacy for the inclusion of science and technology in the school curriculum. The most recent efforts at explicating our conceptions of scientific literacy can be found in the standards and benchmarks work of Project 2061 (American Association for the Advancement of Science, 1990, 1993) and the National Research Council (1996).

The authors' conception of scientific literacy follows from a social responsibility perspective (Waks and Prakash, 1985), that citizens in a global society have an obligation to help resolve the myriad of science- and technology-related societal issues (STS issues for short) that humankind has created through the short-sighted use of science and technology. These include STS issues such as acid rain, enhanced greenhouse effect, ozone layer depletion, ground level ozone pollution, overpopulation, species extinction, water quality and quantity, and waste management. Consistent with this social responsibility perspective and the primacy of scientific literacy as a goal of a school science education, we hold that a scientifically literate citizen is able and willing to take responsible and informed action on STS issues (Rubba and Wiesenmayer, 1997).

The model of instruction we have endorsed for helping learners–citizens gain the knowledge, skills, and willingness to take responsible action on STS issues is known as "STS issue investigation and action instruction." STS issue investigation and action instruction originates from work in environmental education on teaching for responsible citizenship action. That research and more recent research in STS issue investigation and action strategy itself, both of which are summarized elsewhere (Rubba and

Wiesenmayer, 1993), shows that students–citizens continue to take action on societal issues when the instruction helps them develop: (a) an awareness of societal issues, (b) knowledge about actions that might be taken to resolve the issues, (c) the ability to carry out or take informed action on the issues, and (d) certain personality and affective characteristics that dispose one to act (e.g., a somewhat questioning attitude toward technology, an internal locus of control, efficacy perception). STS issue investigation and action instruction incorporates these four critical factors in an integrated four phase structure: foundations, awareness, investigations, and actions.

A unit typically begins with STS issue *foundations* activities in which learners examine the nature of science and technology as well as characteristic interactions among science and technology within society. It is critical that learners understand these interrelationships if they are to take action on an STS issue. Next, in the STS issue *awareness* phase, significant issues facing humankind are identified and analyzed to determine which are "issues" (as opposed to problems) and which are "STS issues" (as opposed to societal issues that might not directly involve science or technology), and to identify related science concepts, technological aspects, social science concepts, and prominent value positions associated with different sides of the STS issue. Case studies delivered in text or video form can be used to develop the next critical understanding: STS issues can and will continue to develop, but will be resolved only through responsible and informed action by citizens.

An STS issue relevant to the community and learners is identified toward the end of the STS issue awareness phase by the class (or a number of STS issues can be selected each by a different group of learners within the class) under the teacher's guidance. This issue should be a derivative of a STS issue with global implications, e.g., acid rain, waste management. It is crucial that learners identify with the issue and that the issue hold potential for learners taking action toward its resolution. The aspect of the issue to be investigated is typically expressed as an STS focusing question (e.g., Does our use of landscape have an impact on global warming?) to provide direction for learners throughout the STS issue investigations and actions phases. A deep understanding of certain science and social science concepts and technological aspects of the issue may need to be developed to help learners clearly define the STS issue. Hence, we encourage a conceptual change approach to teaching in STS.

In the STS issue *investigations* phase, learners develop skills for thoroughly exploring STS issues as they apply those skills in investigating the STS focusing question. These might include learning other science or social studies concepts or aspects of technology that are the foundation for understanding the STS issue through library research using primary and secondary sources, securing data and information from outside agencies, hands-on inquiry activities, collecting natural science data on site, and using social science research techniques such as questionnaires and interviews to collect data within the community. At the close of this phase, the information and data are consolidated by learners to answer the focusing question.

Last, in the STS *actions* phase, learners develop an understanding of various types of action that might be taken in support of the answer formulated to the STS focusing question. A tentative action plan is composed and the pros and cons associated with each action examined from a number of perspectives. Finally, learners decide which action(s) they are willing to take as individuals or as members of a group. They then implement those actions, evaluate the results, and report them to the class.

THE STS LEADERSHIP INSTITUTE

The Leadership Institute in STS Education was funded by a three-year award from the National Science Foundation to develop and cultivate a cadre of science teacher–leaders in STS among the rural middle school–junior high schools within central Pennsylvania and northern West Virginia and to investigate the effectiveness of project-initiated science teacher development activities, as described in greater detail elsewhere (Rubba, Wiesenmayer, Rye and Ditty, 1996). About 24 middle school–junior high school science teachers participated in the institute.

Consistent with its goals, the institute's program of professional development activities was comprised of three summer workshops with follow-up and support activities during the academic years. Global warming was chosen as the STS theme of the project mainly because it is a highly visible STS issue of interest to middle school–junior high school students and lends itself to instruction on science concepts dealt with in life, earth, general, and physical science courses taught at the middle school level, as well as student action. The development and implementation of an STS issue investigation and action unit on three global climate change-related STS issues of enhanced greenhouse effect, ozone layer depletion, and ground level ozone pollution served as the focus for the institute.

As part of an initial summer workshop, six teams of participating teachers developed STS issue investigation and action units on "global warming." Each unit was 60 pages long and included detailed lesson plans for four to five weeks of instruction and had resource materials (e.g., videos, booklets, journal articles, data sheets) referenced or included in appendices. The foundations, awareness, and actions sections of these six units were similar given they were based on a common set of STS outcomes and concepts (Rubba and Wiesenmayer, 1997). However, the investigations lessons were unique to the six units given these included global warming-related science concepts linked to the science course in which the unit was to be integrated.

The results from teacher and student interviews conducted by the authors following implementation of these initial STS units were used by the participating teachers to revise the units during the second summer's workshop. During the third summer and following a second year of implementation, teachers from each of the unit teams and institute staff members met in two writing conferences to merge the individual STS issue investigation and action units into a single unit. The single unit more fully addressed middle school students' alternative conceptions about global warming and ozone depletion, as revealed in the student interviews, and was designed to be used in 6th–9th grade science courses. The unit, "Global Atmospheric Change: Enhanced Greenhouse Effect, Ozone Layer Depletion and Ground Level Ozone Pollution" (Rubba, Wiesenmayer, Rye, McLaren, Sillman, Yorks, Yukish, Ditty, Morphew, Bradford, Dorough, and Borza, 1995) can be found on-line at http://www.ed.psu.edu/CI/Papers/sts/gac-main.html.

THE SCIENCE OF GLOBAL WARMING

The IPCC is considered the world's foremost scientific program on the assessment of climate change (Kerr, 1997). IPCC Working Groups I, II, and III are charged with assessing, respectively, the underlying science, impacts and response options, and economic and social ramifications relative to climate change. The reports emanated from these working groups provide a thorough and timely account of the science that underlies climate change (IPCC publications are listed at http://www.ipcc.ch; summaries of reports can be viewed on-line). The first report of IPCC Working Group I, *Climate Change: The IPCC Scientific Assessment* (Houghton, Jenkins, and Ephraums, 1990), was published in 1990. It was followed by two interim reports in 1992 (Houghton, Callander, and Varney) and 1994 (Houghton, Meira, Filho, Bruce, Lee, Callander, Haites, Harris, and Maskell), the former coinciding with the 1992 "Earth Summit" in Rio de Janeiro and both adding to our understanding about the role of atmospheric aerosols and, more broadly, the driving phenomenon of radiative forcing, in climate change. The most recent IPCC Working Group I report is *Climate Change 1995: The Science of Climate Change* (Houghton et al., 1996).

Much of "the science" of global warming, as set forth in this section, is based on this latest IPCC report as well as the second edition of *Our*

Changing Planet by Mackenzie (1998). What is provided herein only summarizes some basic information from this extensive body of knowledge. Consider this excerpt from the preface of *Climate Change 1995*, which speaks to the validity and serious nature of the first IPCC scientific assessment as published eight years ago (Houghton *et al.*, 1990):

> We believe the essential message of this report continues to be that the basic understanding of climate change and the human role therein, as expressed in the 1990 report, still holds: Carbon dioxide remains the most important contributor to anthropogenic forcing of climate change; projections of future global mean temperature change and sea level rise confirm the potential for human activities to alter the Earth's climzate to an extent unprecedented in human history; and the long time-scales governing both the accumulation of greenhouse gases in the atmosphere and the response of the climate system to those accumulations means that many important aspects of climate change are effectively irreversible. (Houghton *et al.*, 1996, p. xi)

While the mechanisms and feedback loops that underlie global warming are multiple, interrelated, and sophisticated (Houghton *et al.*, 1992, 1996; Mackenzie and Mackenzie, 1995; Mackenzie, 1998; National Academy of Sciences, 1992), global warming ultimately is a natural phenomenon involving sunlight, or more precisely, solar radiation incident upon the earth and the atmosphere. Without some degree of global warming—what has been more commonly referred to as the "greenhouse effect" (even though the mechanism is not congruent with the way greenhouses are heated by solar radiation)—the earth would be a frozen planet, or at the very least, temperatures at earth's surface would not be the average 15°C (excludes Antarctica) we know and the diurnal variance in temperature would be considerably greater (Mackenzie, 1998). The mechanism involves energy from the sun—ultraviolet, visible, and infrared wavelengths of radiation in the electromagnetic spectrum—being absorbed by the earth's surface. This energy is re-emitted by the earth's surface as infrared radiation, which is trapped and re-radiated by "greenhouse" gases present in the lower atmosphere (troposphere). Accordingly, the "greenhouse effect" (necessarily) warms earth's surface temperature and is a major factor in making life possible.

Foremost among these gases are carbon dioxide, methane, nitrous oxide, and water vapor, all of which have natural origins, along with CFCs (chlorofluorocarbons), which were recently introduced by humans (CFCs are exclusively of anthropogenic origin). Changes in the atmospheric level of these greenhouse gases, especially carbon dioxide, over the past 160,000 years have corresponded to changes in global temperatures (O'Meara, 1997).

The natural greenhouse effect is being "enhanced" from anthropogenic sources of carbon dioxide, methane, nitrous oxide, water vapor, and tropospheric ozone: Atmospheric levels of these gases are increasing due to industrialization of the world. This is producing a positive radiative forcing (the "enhanced" greenhouse effect) in the earth's atmosphere system and giving rise to global warming. Among all greenhouse gases, the accumulation of carbon dioxide makes the largest contribution to global warming. However, CFCs add further to the accumulation of greenhouse gases in our atmosphere, and as such, to this positive forcing. CFCs were developed for use as aerosol propellants, cleaners, foam-blowing agents, refrigerants, and coolants. CFCs are powerful greenhouse gases that have long atmospheric lifetimes (50–100 years), during which they not only act as greenhouse gases, but also are responsible for ozone layer depletion—two different environmental problems that can be easily confused. Pursuant to the international agreements set forth in the Montreal Protocol and subsequent amendments (Mackenzie, 1998), the atmospheric growth rates of CFCs are slowing (World Meteorological Organization, 1995). However, at the same time, levels of CFC substitutes (e.g., hydrochlorofluorocarbons [HCFCs] and hydrofluorocarbons [HFCs]) are increasing in our atmosphere. Although the atmospheric lifetimes and ozone depletion potentials of these CFC substitutes generally are much less than CFCs, they are still "notable" greenhouse gases (World Meteorological Organization, 1995).

Metzger (1996) acknowledges the "daunting" task of pulling together some basic information on global atmospheric change for precollege teachers. She suggests related student projects, including one that engages students in learning about CFCs and their role in global warming and ozone layer depletion. The interrelationship between ozone layer depletion and global warming is one of the best examples of the complex nature of global atmospheric change (Mackenzie, 1998). The material that follows attempts to illustrate that complexity.

The most recent IPCC report (Houghton et al., 1996) explains that CFCs, as a greenhouse gas in the troposphere, provide a positive feedback (radiative forcing) to potential global warming. However, because CFCs provide a *negative* feedback to global warming through stratospheric ozone destruction, their net radiative forcing (or global warming potential) is reduced. Mackenzie (1998) defines a negative feedback as "a moderating effect on an initial disturbance" (p. 369)—that disturbance in this case is a reduced ability of the earth–atmosphere system to "cool to space" (Houghton, 1996, p. 14) because of the accumulation of heat-absorbing greenhouse gases in the troposphere. The 1995 IPCC report reaffirms and extends the understanding about CFCs' dual and opposing contributions to global warming that was set forth previously by the IPCC (Houghton *et al.*, 1992; Houghton *et al.*, 1994; World Meteorological Organization, 1995).

With the forecasted decline in stratospheric CFC levels early in the next century, it follows that "the [resulting] ozone recovery constitutes a *positive* radiative forcing that acts to enhance the effect of the well-mixed greenhouse gases" (Sanhueza and Zhou, 1996, p. 110). A knowledge of the "greenhouse" properties of ozone and the vertical location in the stratosphere of ozone depletion and recovery (e.g., depletion in lower stratosphere gives rise to a negative feedback to global warming) is important to understanding feedbacks to global warming consequent to changes in the quantity of stratospheric ozone (Houghton *et al.*, 1992; Solomon and Srinivasan, 1996).

The complexity of this issue is magnified by other scientific and media reports the public, including teachers and students, encounter that do not mention these opposing actions of CFCs in global warming and instead state that ozone layer depletion may magnify global warming. For example: "Ozone absorbs most of the sun's harmful rays; its loss intensifies the greenhouse effect" ("U.N.: Ozone thinner than ever," 4/9/97, p. 5A). Some of these reports, such as "Stratospheric ozone depletion" (United States Environmental Protection Agency, 1995), attribute this intensification to potential reductions in photosynthesis due to increases in ultraviolet-B radiation (UV-B). UV-B may decrease phytoplankton and green plants, which will reduce the biosphere's sink capacity for carbon dioxide, thereby further increasing the atmospheric level of this greenhouse gas.

Others (e.g., Denman, Hofmann, and Marchant, 1996) believe there is considerable uncertainty about the effect of phytoplankton reductions on global warming and note a need for further research. Another factor related to the biological activity of phytoplankton is that it produces dimethylsulfide (DMS) gas, which is subsequently released to the atmosphere from the ocean surface (Mackenzie, 1998). DMS leads to the formation of sulfate aerosols, which in turn (as cloud condensation nuclei) give rise to clouds. Both sulfate aerosols and clouds reflect incoming solar radiation back to space, and accordingly, provide a negative feedback to global warming. Therefore, reductions in phytoplankton could lead to less reflection of solar radiation, which would add to global warming. Additionally, there is speculation that global warming will contribute a positive feedback to ozone layer depletion. The mechanism of action here is that global warming traps heat in the troposphere and creates a colder stratosphere. Colder temperatures catalyze chlorine-induced destruction of the ozone layer (Austin, Butchart, and Shine, 1992).

Accordingly, teaching about the role of CFCs in global climate change and related STS issues can be very challenging to the science teacher.

Adding to this task, as is clearly noted below, upper elementary through secondary level students may bring to or formulate during instruction the alternative concept that the ozone hole causes or enhances the greenhouse effect because it lets through more sunlight or UV-B, warming the planet (Boyes and Stanisstreet, 1993; Dorough, Rubba, and Rye, 1995; Christidou and Koulaidis, 1996).

Studies on Student Pre-Instructional Understanding of Global Warming

Several studies (Boyes and Stanisstreet, 1993; Christidou, 1994; Christidou and Koulaidis, 1996; Dorough et al., 1995; Francis, Boyes, Qualter, and Stanisstreet, 1993; Koulaidis and Christidou, 1993; Plunkett and Skamp, 1994; Potts, Stanisstreet, and Boyes, 1996; Rye, 1995) referred to below have investigated upper elementary and secondary school students' understandings of concepts related to global warming and (stratospheric) ozone layer depletion. These studies either state explicitly that they are pre-instructional in nature or they lack any specific instruction.

Christidou (1994) conducted a series of three interviews with 41 5th–6th grade students to investigate how they processed information about ozone layer depletion and the greenhouse effect. Approximately 63 percent of the students believed that increased UV consequent to ozone layer depletion would warm up the earth and melt polar ice caps. Some students referred to ozone layer depletion as the primary cause of the greenhouse effect. Overall, the findings suggested that students had greater familiarity with ozone layer depletion than the greenhouse effect and that students often conceptualized these two phenomena as one.

Francis et al. (1993) administered a questionnaire about the greenhouse effect to 565 children aged 8–11 years, and interviewed a small subset (15) of these individuals. In these interviews, many students introduced the phenomenon of ozone layer depletion. Various students' responses suggested that they fused, and as such confused, the ideas of global warming and ozone layer depletion. The authors conclude that such may be "embedded in an apparently logical conceptual framework: the idea that damage to the ozone layer allows the 'sunshine' to penetrate and so warm the earth" (p. 390).

Boyes and Stannisstreet (1993) reached conclusions similar to those of Francis et al. (1993) from their study that investigated understandings about the greenhouse effect in 861 students aged 11–16 years. They concluded that students link CFCs primarily to ozone layer depletion and that they blend the ideas of global warming and ozone layer depletion: "[T]here is a likelihood of pupils reaching the correct conclusion via an erroneous pathway, in that an affirmation that CFCs affect global warming may arise from a knowledge of the CFC-ozone connection and a confusion between the two global environmental effects" (p. 550).

Plunkett and Skamp (1994) interviewed 45 4th–8th grade students about the ozone layer and ozone hole. They found that about 20 percent of students believed that aerosol sprays destroyed the ozone layer and over 25 percent believed the ozone layer hole would lead to climatic changes that included the melting of polar ice caps. These authors state as the main conclusion of their study that students have a conceptual framework that confuses ozone layer depletion and the greenhouse effect.

Dorough *et al.* (1995) investigated the pre-instructional understandings about global warming and ozone among 22 5th–6th grade students. About 40 percent of the students introduced the concept of ozone or ozone layer in response to the query: "When you think about global warming, what thoughts come to mind?" Several of the students gave evidence of believing that ozone layer depletion is a major contributor to global warming. Few students gave evidence of knowing about CFCs or the greenhouse effect.

Rye (1995, 1998) interviewed 38 grade eight physical science students about CFCs and their role in global atmospheric change prior to instruction from lessons (Rubba, Wiesenmayer, Rye *et al.*, 1995) that investigated the "science" of global atmospheric change. Over half of the students were unfamiliar with CFCs. Of those who did appear to hold scientific understandings about CFCs, some acknowledged that they were only guessing and others gave evidence of holding alternative conceptions about the sources of CFCs or interrelationships between ozone layer depletion and global warming. For example, approximately 30 percent of the students did (appropriately) connect CFCs to destruction of the ozone layer. However, some of these students believed that CFCs came from the combustion of fossil fuels or that they were a problem because they would allow more of the sun's rays (UV rays specifically) to hit earth, melting glaciers and ice caps, causing flooding, and so on. Additionally, some of these students inferred that ozone depletion caused an elevation in temperature on earth (however few actually labeled the latter as "global warming"). Rye's findings about ultraviolet rays heating up earth and melting ice caps also surfaced in the studies of 11–13-year-old students by Christidou and Koulaidis (1996) and Potts *et al.* (1996). For example, the former report that "[Q]uite often children attributed thermal properties to ultraviolet rays" (Christidou and Koulaidis, 1996, p. 434).

A common finding to each of these studies is that prior to instruction students may hold, to an appreciable degree, alternative conceptions (Wandersee, Mintzes, and Novak, 1994) about global warming and its association with ozone layer depletion. These alternative conceptions, of course, differ from expert scientific knowledge (Abimbola, 1988; Houghton *et al.*, 1992, 1995), as revealed through the previous section of this chapter on the science of global warming: "[T]heir [CFCs and HCFCs] net radiative forcing is reduced because they have caused stratospheric ozone depletion which gives rise to a negative radiative forcing." (Houghton *et al.*, 1996, p. 3). Indeed, Mackenzie (1998) reveals that depletion of ozone in the stratosphere offset approximately 20 percent of the total greenhouse warming between 1980 and 1990. These conclusions are based on radiative balance calculations and account for the fact that, "Ozone is an effective greenhouse gas both in the troposphere and the stratosphere" (Houghton *et al.*, 1992, p. 8). Accordingly, a decrease in the amount of stratospheric ozone means that less (as opposed to more) infrared radiation will be radiated back toward earth.

Student Post-Instructional Understanding of Global Warming

Procedures. As was noted previously, during the first academic year of the project, six STS issue investigation and action units on global warming were developed, each by a team of the teacher-participants. These units shared a core of content on the nature, cause, and resolution of global warming. The six STS global warming units were field-tested by each of the teachers on the respective teams during the first academic year. To inform the unit revision, the project staff interviewed the teachers and a sample of their students approximately two weeks following the end of each STS global warming unit. The unique purpose of the student interviews was to investigate their understandings relative to STS global warming unit content, ascertain what citizenship actions they had taken toward the resolution of global warming, and learn their perceptions of the importance as well as the strengths and limitations of the STS global warming unit. We report herein only on findings related to the understanding of global warming held by a sample of 24 6th–8th grade students who had completed one of four STS global warming units. Relevant, select findings from the teacher interviews are included here. Details on the research procedures used in interviewing the students and teachers are provided elsewhere (Rye, Rubba, and Wiesenmayer, 1997).

A open-ended interview protocol (Patton, 1987) was developed by the authors and field-tested (Osborne and Freyberg, 1985; Novak and Gowin, 1984). As can be implied, the sample for this study was one of convenience, predominantly white and drawn from students in four middle level grade classrooms (two from 6th grade and one each from 7th and 8th grades). These students collectively represented instruction from four of the six STS global warming units. (The other two units were not included in this study due to the high school level focus of one unit and teacher–participant attrition associated with the other unit.) The interview questions focused on eliciting, sequentially, students' understandings and views in the following areas: (a) the nature and cause of global warming; (b) what global warming unit content was "important"; (c) why global warming is an STS issue; (d) possible citizenship actions to resolve global warming; (e) actions actually taken to help resolve global warming; (f) likes and dislikes about the global warming unit; and (g) connections between global warming and ozone. The questions to elicit connections students perceived between global warming and ozone were placed at the end of the interview.

Transcripts of student interviews were examined to reveal evidence of understanding of global warming using an "expert" concept map as a template. Figure 1 presents the "expert" concept map, which sets forth in a hierarchical conceptual network (Heinze-Fry and Novak, 1990; Jonassen, Beissner, and Yacci, 1993; Lomask, Baron, and Grieg, 1993; Novak and Gowin, 1984) core content on the nature, causation, and resolution of global warming that was shared by the STS global warming units. The concept map was used to assess the degree to which students gave evidence during the interview of holding scientifically appropriate concepts and concept relationships present in the units.

Analysis of the interview data was guided by these two questions:

- What post-instructional alternative conceptions do students hold of the nature, cause, and resolution of global warming?
- In what ways and to what extent do these alternative conceptions incorporate connections with ozone layer depletion?

Assertions were formulated and validated in regards to alternative concepts students held about the nature, cause, and resolution of global warming. Specifically, researcher assertions of the presence of such conceptions, based on single or multiple instances in the transcript, were verified by studying the entire transcript for the presence of responses that would validate further or invalidate the assertions. Principles of inductive and logical analyses (Patton, 1990) were employed to a limited extent in order to label and combine into broader categories the emergent alternative conceptions. Fictitious names have been assigned to all students and teachers for whom transcript excerpts have been included.

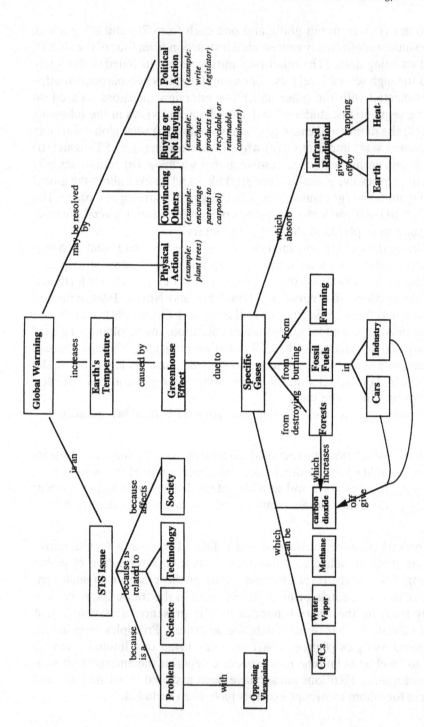

Figure 1. Expert concept map of global warming. *Note:* From "An investigation of middle school students' alternative conceptions of global warming," by J. Rye, P. Rubba, and R. Wiesenmayer, 1997, *International Journal of Science Education* 19:532. Copyright 1997 by Taylor and Francis Ltd. Reprinted with permission.

Findings and Discussion

Tammy: Yeah, I think it's basically, that everything works against the ozone and when the ozone breaks, that's when the global warming is going to come.[1]

Candy: [We should] start buying things that like don't have those kind of gases in them. Like instead of buying hair spray with CFCs, you can buy some like in a pump or something.

Billy: ... CFCs cause destruction of the ozone layer to let the sun, the ultraviolet rays, get in and heat up the earth.

Evert: ... And all those oil wells [on fire in the Middle East] gave off tons and tons of carbon dioxide. Which made the ozone layer just come apart and that's where the biggest hole came from.

Sally: It [carbon dioxide] comes out of the exhaust. And as it goes up to the ozone past the stratosphere. Then after it hits the ozone it like eats it up and then UV come and hit earth.

These transcript excerpts, taken from interviews with several students, speak to the principal alternative conceptions found to be held by the students. These alternative conceptions, each connecting ozone layer depletion and global warming in some way, were as follows:

- Ozone layer depletion is a principal cause of global warming (Tammy)
- Aerosol sprays contain CFCs and damage the ozone layer (Candy)
- CFCs cause global warming exclusively through destroying the ozone layer (Billy)
- Carbon dioxide destroys the ozone layer (Evert)
- Carbon dioxide causes global warming exclusively by destroying the ozone layer (Sally)

Even though they are limited by the small sample sizes and a volunteer sample (Roberts, 1992), no statistically significant correlations (phi coefficients) were found between overall academic ability level or gender and the degree to which any of these alternative conceptions were held among the 24 students interviewed. Additionally, we did not find significant correlations (point biserial) between the degree to which students held a

---

[1] This and all subsequent transcript excerpts used in this chapter are from "An investigation of middle school students' alternative conceptions of global warming," by J. Rye, P. Rubba, and R. Wiesenmayer, 1997, *International Journal of Science Education*, 19, p. 536, 538–544, and 546–547. Copyright 1997 by Taylor & Francis Ltd. Reprinted with permission.

specific scientific understanding about global warming—the score derived by assessing their interview transcript against the expert concept map (Fig. 1)—and the prevalence of these alternative conceptions.

The prevalence in students of these alternative conceptions did not appear to differ appreciably across the STS global warming units, although students who were the recipients of instruction from one of the units gave little evidence of believing that carbon dioxide destroyed the ozone and did not appear to believe that the exclusive role of CFCs in global warming was via ozone layer destruction. However, this study was not designed to ferret out differences between students according to the unit of instruction completed. Below we examine in some detail our findings relative to each of the five alternative conceptions. However, this separation is somewhat artificial given these alternative conceptions are interwoven in students' cognitive structures and the idea that "ozone layer depletion is a major cause of global warming" surfaces continuously.

Ozone Layer Depletion as Principal Cause of Global Warming. Seventy-five percent ($n = 18$) of the students and the majority within each academic ability level believed that ozone layer depletion or increased UV radiation helped cause global warming. More important, 54 percent ($n = 13$) of all students gave evidence of believing that ozone layer depletion, or consequent increases in UV or the sun's rays, was a major or predominant cause of global warming. Students' supporting reasons frequently were related to natural phenomena; for example, the heat from the increased sunlight heats up the planet. Example transcript excerpts follow that illustrate how students connected ozone and global warming.

Julie suggested that ozone depletion allows sunlight to penetrate the atmosphere, warming the earth.

> Interviewer: Okay, is there anything else that comes to mind when you think about global warming?
> Julie: Heat.
> Interviewer: Heat. How does heat come to mind?
> Julie: Umm . . . Well, destroying the ozone layer and the sun making it warmer.

Sally had this to say, when asked what ideas came to mind when she "thought about global warming":

> Sally: Ah, the ozone and how it is being affected by CFCs, umm, carbon dioxide, and stuff that can cause the ozone to deteriorate. . . .
> Interviewer: Okay, what makes you think about ozone?

Sally: Umm, mainly stuff that can harm the ozone layer. Like the CFCs and the air conditioners in cars and when the cars become old they leak. . . .

Interviewer: Okay, and how does it harm the ozone?

Sally: Well, there's like chloro-fluoro-carbons that go up and eats the like ozone up. And causes the ozone hole to get bigger and bigger and then UV rays from the sun enter and hit earth. . . .

Interviewer: Umm, and how is that, how is that related to global warming though?

Sally: Well, the UV, umm, rays hit earth and earth will get warmer. And that's what causes global warming.

Later in the interview, the researcher again queried Sally about the ozone hole and global warming:

Interviewer: How is that [ozone] hole related to global warming?

Sally: Well, what it does is, like I said, the Sun's UV rays come through the ozone. And then when it hits earth, the earth will warm. And that's how you basically get global warming.

Questions were placed at the end of the interview about global warming to further elucidate students' conceptions about connections between global warming and ozone layer depletion. Most students who believed ozone layer depletion was a major cause of global warming had already provided substantial evidence of this alternative conception by the time these questions were introduced. Still, student responses were useful in confirming this and other researcher-held assertions. The following transcript excerpts for Charles, Evert, and Barton illustrate these student responses to the initial general query made by the interviewer, "When you think about ozone, do you also think about global warming?"

Charles: Well, a little bit because CFCs destroy the ozone. . . . UV rays are basically what global warming is when they come in, you know. . . . And it worries me that they could come in faster and faster and then they could bounce off faster and get hotter and hotter and hotter.

Evert: Well, yeah because the ozone is what, I think, is going to keep global warming out.

Interviewer: And why do you think about global warming when you think about ozone?

Barton: Because the global warming affects the ozone, it'll burn a hole in it.

As evidenced by the last transcript excerpt, Barton also believed that global warming causes ozone depletion. Candy believed the greenhouse effect causes ozone depletion:

Candy: . . . [T]he greenhouse effect is when the gases and that are ruining, causing holes in the ozone which let off UV rays.

The idea that global warming or the greenhouse effect causes ozone layer depletion may have been held by other students given a number of students mentioned that ozone layer depletion (or increased UV radiation) "comes to mind" when thinking about global warming, and carbon dioxide destroys the ozone layer. This "construction" also could be the result of a surface understanding and inappropriate logic regarding the role of CFCs in both global warming and ozone layer depletion. That is, if CFCs cause global warming and destroy the ozone layer, then global warming destroys the ozone.

Barton suggests that aerosol sprays containing CFCs play a central role in global warming mediated destruction of the ozone layer.

Barton: Global warming, ah, it burns holes in the ozone layer and it can do that by using hair spray and stuff like that. Like aerosol sprays and stuff like that. . . . Hair spray has CFCs in it. And if you use enough of that it will burn a hole in the ozone layer. . . .

This idea surfaced many times in his interview and is explored more fully here.

Aerosol Sprays Contain CFCs and Destroy Ozone. The idea that aerosol sprays contain CFCs and destroy the ozone layer was an alternative conception found to be prevalent in 54 percent ($n = 13$) of the students. Globally, aerosol sprays account for some CFC use—about 15 percent in 1993 (Kurtis Productions, 1993). However, the sale of aerosol cans containing CFCs was banned in the United States and Canada in 1979 (University Corporation for Atmospheric Research, 1992). Hence, aerosol cans sold in the United States should not contain CFCs.

Mason: Ah, well when you use hair spray. Anything that's in a canister like hair spray and cooking spray for your pans and stuff has CFCs so it, umm . . . will shoot out or compress air. And then it's CFC. And that stands for chlorofluorocarbons, and they eat away at the ozone.

The idea that, in the United States, aerosol cans—hair spray in particular—contain CFCs and, thereby, destroy the ozone layer is a prevalent and perhaps tenacious alternative conception (Kurtis Productions, 1993). Barton's interview reveals the great extent to which he appeared to hold this alternative conception:

Interviewer: ... What is an important thing you learned from the unit on global warming?

Barton: The most important thing I learned is, how to prevent global warming from happening by, like using filters and stuff like that. And using hair sprays that are like a pump.

Interviewer: Okay. What can be done about global warming?

Barton: ... You just have to kind of watch the products that you use. ... Hair spray ... Um, I imagine some deodorants would have CFCs in them.

This interview excerpt for Barton suggests that alternative conceptions can result in misdirected citizenship actions toward issue resolution. In response to the interview questions about the resolution of global warming, the majority of students holding the alternative conception that aerosols contain CFCs and destroy the ozone stated actions targeted at decreasing the use of aerosols. Examples of such actions can be found in the transcript excerpt of the interviews with Candy and Barton (previously presented).

This alternative conception may have been developed or reinforced in some students by instruction from the units. Some STS global warming units contained content stating that aerosol use has been a factor in ozone layer depletion. This contention was supported when one of the institute teacher-participants noted that she "stood corrected" on CFC use and had been teaching students that aerosols sold in the United States at the present time contained CFCs. However, this teacher was not one of the instructors of the students who participated in this study.

Exclusive Role of CFCs in Global Warming. In response to interview questions specifically targeting student understanding of the nature, cause, and resolution of global warming, 50 percent ($n = 12$) of the students provided evidence of the scientifically accurate understanding that CFCs are a potential cause of global warming. However, approximately half of those students ignored CFCs' central mechanism as a heat trap in global warming (Mitchell, 1989) and, in fact, held the belief that the exclusive role of CFCs in the causation of global warming was through depleting the ozone layer. Several of the transcript excerpts presented previously suggested that certain students may hold this alternative conception. As an example, further questioning of Julie on the role of CFCs in global warming did not reveal an understanding of CFCs' "greenhouse gas" property. However, it provided further evidence that her conception of CFCs' mechanism in global warming was "ozone layer depletion."

Interviewer: Are they [CFCs] connected to global warming?
Julie: Yes, I think.... Umm ... Well, CFCs destroy the ozone layer and that, I think, that's what happened that there is a hole in it from the CFCs.
Interviewer: Okay. Is there another important thing you learned from the unit?
Julie: Umm, that CFCs are extremely harmful. Like before no one really worried about them and never thought about them. But now as soon as like you spray something it is in the air and ...
Interviewer: Okay. And why is it important for you to know about the CFCs?
Julie: Well, like I said before, it eats the ozone layer and the sun gets much hotter and everyone will be so hot.

Students who held the alternative conception that CFCs cause global warming exclusively via ozone layer depletion also were likely to believe that carbon dioxide (or exhaust, smoke, or car pollution) causes global warming exclusively by destroying the ozone layer. A significant correlation (phi coefficient of $r = .41, p < .05$) was found between students holding these two alternative conceptions. The idea that carbon dioxide causes global warming exclusively through ozone layer depletion was a subset of the more widely held alternative conception that carbon dioxide destroys the ozone layer.

Carbon Dioxide Destroys the Ozone Layer. Over 70 percent ($n = 17$) of the students provided evidence of understanding that carbon dioxide is a potential cause of global warming when they were asked about the nature, cause, and resolution of global warming. However, "ozone layer destruction" was sometimes cited as one possible or the mechanism for how carbon dioxide caused global warming, as is evident in Mason's response to a probe by the researcher.

Interviewer: How are cars a cause of global warming?
Mason: When you use them a lot. The gas—I think it's carbon monoxide and dioxide from your exhaust—form what's called a bad ozone, and that's smog around the lower part of the earth, or make holes in the good ozone that surrounds the earth.

Students sometimes referred to "exhaust," "smoke," or "car pollution," rather than "carbon dioxide" destroying the ozone layer. When such instances were collapsed, 50 percent ($n = 12$) of the students were found to hold the belief that carbon dioxide (or exhaust, etc.) destroys the ozone layer. Students who spoke of exhaust (etc.) as an ozone destroyer may have been thinking about an ingredient in exhaust (etc.) other than, or in addition to, carbon dioxide. For example, one student stated she thought car pollution "contains CFCs and things that help destroy the ozone layer."

Still, some of the students holding the alternative conception that carbon dioxide (or exhaust, etc.) destroyed the ozone layer, also provided a "scientifically acceptable" mechanism of action of carbon dioxide in global warming. As Samuel did, they noted that carbon dioxide traps heat in the lower atmosphere (Mitchell, 1989).

Interviewer: ... Any other causes [of global warming] come to mind?

Samuel: Umm. Methane, carbon dioxide are some of the gases that break down the ozone layer. And there's a bad ozone and, ah, nitric oxide.

Interviewer: Okay, is there anything else about carbon dioxide and methane that make them a cause of global warming?

Samuel: They keep the heat in.

A third of all students gave evidence of believing that the exclusive role of carbon dioxide (or exhaust, etc.) in the causation of global warming was via destroying the ozone layer, as is exemplified here by an interview excerpt from Andrew.

Interviewer: Okay, carbon dioxide. Whatever you see as being the causes [of global warming], that's what I'd be interested in knowing about.

Andrew: That's like, when the carbon dioxide is let off and it breaks down the ozone layer, it lets more sunlight in and then that makes it get warmer. That's what I think is causing it.

Interviewer: Are there other ways that carbon dioxide would be related to global warming?

Andrew: Umm ... No, I don't think there is.

Some of the students may have held this alternative conception because they perceived carbon dioxide as being similar to CFCs in that both CFCs and carbon dioxide contain the word carbon, both are greenhouse gases, and both are cited as components of pollution. Further, several students lumped together various greenhouse gases as agents that deplete the ozone layer, and they spoke collectively about carbons and pollution. One student put it this way: "It's sort of like the gases that trap the sun rays also eat the ozone layer away." Other examples of this thinking can be found in excerpts shown previously from the interviews with Sally and Samuel.

Most (75 percent) of the students who believed carbon dioxide (or exhaust, etc.) destroys the ozone layer, and all of the students who believed this was the exclusive role by which it caused global warming also held the alternative conception that ozone layer depletion is a major cause of global warming. A significant correlation (phi coefficient of $r = .65$, $p < .01$) was found between students holding the latter two alternative conceptions. These findings warrant further investigation. The STS global warming units emphasized that carbon dioxide could be a major cause of global warming. Perhaps these students concluded that, since carbon dioxide destroys the ozone layer, it follows that ozone layer depletion must be a major cause of global warming.

Few students provided a scientifically complete (relative to the expert concept map presented in Figure 1) understanding of the radiative mechanism of greenhouse gases in the causation of global warming. Students were more likely to note that various gases (or pollutants or car exhaust) trapped UV or the sun's rays, thereby causing global warming, as opposed to trapping or absorbing infrared radiation given off by the earth.

The former belief is relative to the cause of global warming and may reflect an unclear understanding of the role of the different types of radiation in global warming. Nonetheless, it apparently is also related to the students' conceptions of the connections between global warming and ozone layer depletion. For, in our study, even the students who gave evidence of a relatively sound understanding of the radiative mechanisms underlying global warming (greenhouse gases absorb infrared or long wave radiation emitted by earth that originated from shorter wavelength radiation from the sun, thereby trapping heat) were likely to see the consequences of both global warming and ozone layer depletion as warming the earth. Mollie's thoughts, as verbalized in the interview, provide an example of this:

Interviewer: Okay, when you think about ozone do you think about global warming?

Mollie: Yeah. Because sometimes I do ... And it [earth] also gets like warmer because of there's a hole in the ozone.

Then the rays are going to come in at like a hotter temperature.

Interviewer: Okay. What is it about ozone that makes you think about global warming?

Mollie: Like the, how the sun rays come in and make the earth like warmer and like hotter. That like happens in both, so that's what makes it, I think.

To facilitate the focus and clarity of instruction (Wise and Okey, 1983), our teacher-participants were advised to not include instruction on ozone in their initial STS global warming unit. However, an inspection of these units suggested that three of the four teacher-participants (of students who participated in this study) did address ozone layer depletion during STS global warming instruction. This instruction may have helped develop or reinforce the alternative conceptions that surfaced in this study. Post-instructional interviews with these four teachers revealed that they believed some degree of confusion between global warming and ozone layer might exist among the students. The following are select responses from the four teachers to the related interview question: "Because we are going to ask your students, or we have asked your students, about ozone, we need to know how you dealt with ozone?"

Sue: Good or bad ozone? ... We talked about both and that was a hard thing for them to understand. ... I used the diagram, two different diagrams, that I had in the units to show them, ah, greenhouse effect, and what part the ozone played as well.

Mort: And I hope they know the fact is that it's [ozone depletion is] not related to global warming. I don't know if all of them will know that.

Jack: As a separate problem that really had no connection with global warming. ... There was one other problem I had. And it was separating in their minds, ah, the problem with ozone depletion and the problem with global warming. ... And then there were other articles [that the librarian put together] that actually drew a connection and it was the first time I had heard that the fact that the ozone is being depleted also is a negative feedback for global warming. Is that right? ... That ... confused me.

Mary: ... I don't feel we did a good job at that. ... [A]nd I could not avoid the ozone issue so I'm sure I don't know what your questions are for the students, but they might have some confusion. ... [W]e just said try to keep them separate.

DISCUSSION

These interview findings suggest that middle school students may hold specific alternative conceptions about global warming that limit and confound their understanding of its nature, causation, and potential resolution. The studies by Boyes and Stanisstreet (1993), Christidou (1994), Dorough *et al.* (1995), Francis *et al.* (1993), Plunkett and Skamp (1994) and Rye (1995, 1998) reviewed earlier on students' understandings related to global atmospheric change suggest that students likely entered instruction with a greater awareness of the ozone layer than of global warming, and that some students may begin instruction with the alternative conception that the "hole" in the ozone layer causes global warming. Given that over half of the students in our (post-instructional) study believed that ozone layer depletion was a major cause of global warming, this alternative conception may be tenacious.

Knowing the pre-instructional conceptions of the students interviewed in this study likely would have proven useful in determining alternative conceptions that appear to develop in students as a result of formal instruction, and what pre-instructional understandings appear to interact with formal instruction to yield those unintended learning outcomes (Wandersee *et al.*, 1994). However, limitations in working with the schools and teachers did not allow that. Still, the investigation of pre-instructional understandings about global warming and ozone among 5th and 6th grade students conducted by Dorough *et al.* (1995) involved students from schools similar to those in which the data reported here were collected.

Dorough's interviews revealed that "ozone layer" came to mind in more than a third of the students as they "thought about global warming." Francis *et al.* (1993) also reported that students often introduced, during interviews about the greenhouse effect, the concept of the ozone layer. Initial exposure to the concept of global warming may evoke thoughts in students about the ozone layer.

Even though global warming is a sophisticated concept (Houghton *et al.*, 1992; Mackenzie and Mackenzie, 1995; National Academy of Sciences, 1992), learners will bring to global warming instruction the intuitive knowledge (West and Pines, 1985) that the sun feels warm and that a sunburn makes us hot. Such intuitive knowledge can interact with new information to yield unintended learning outcomes, e.g., the alternative conception that the extra sunlight or ultraviolet radiation coming through the "hole" in the ozone layer heats up the planet.

Additionally, the "greenhouse effect" is part of the public vocabulary (Ennis and Marcus, 1994; Pomerance, 1989). Many students likely have heard the term prior to related formal instruction (Hocking *et al.*, 1990). We speculate that concepts such as "ozone hole," "UV rays," "CFCs," and "greenhouse effect" may be "loose" in many students' cognitive structures and connected inappropriately to make sense of formal instruction on global warming. Furthermore, there is real potential for student confusion and construction of alternative conceptions when they are confronted with multiple phenomena that are explained through some of the same concepts. Transcript excerpts from the interview with Candy are illustrative:

> Interviewer: Okay. Let's say you were to explain global warming to someone else. What exactly is global warming? How would you explain it?
>
> Candy: It's, well, I don't know. Let me think . . . Umm . . . It's when like CFCs and things. Or it's like greenhouse. Well, I kind of get, I kind of mix them together cause that's what you think about them. They're kind of, people like talk about them as the same thing. But they're not.
>
> Interviewer: Sounds like you're unsure of something.
>
> Candy: Yeah, I'm getting confused.
>
> Interviewer: Tell me what you're unsure of.
>
> Candy: I guess, it must be when I speak about global warming or greenhouse effect. You get the idea that they're exactly the same thing. But they're two different subjects and then I get confused . . . cause I know greenhouse effect is ozone depletion like. . . . Because both things [global warming and ozone] are like more heat or something getting into the earth, which kind of confuses you sometimes when you try to think of which is which.

In the design and delivery of global warming instruction, concepts and propositions embedded in our findings of alternative conceptions need greater attention. When instruction attempts to present both global warming and ozone layer depletion within the same instructional unit (Koulaidis and Christidou, 1993), learners may formulate the erroneous idea that ozone layer depletion is the principal cause of global warming due to the involvement of incoming solar radiation and CFC greenhouse gases in both environmental problems (Hocking, Sneider, Erickson, and Golden, 1990; Monastersky, 1992; Pomerance, 1989). Worrest, Smythe, and Tait (1989) recommend addressing ozone layer depletion and global climate change together. However, because the interrelationships between changes in atmospheric ozone and global warming are so complex, many researchers (Ashmore and Bell, 1991; Fishman, 1991; Houghton *et al.*, 1992; Lacis,

Wuebbles, and Logan, 1990) believe the development of scientifically appropriate conceptions of global warming may be confounded by integrating the study of both topics. Conversely, because global warming and ozone depletion have received considerable attention by the media, a superficial coverage (or no coverage) of ozone depletion may result in the development of alternative conceptions about global warming.

Our current position is that it is likely impractical to avoid the topic of ozone when teaching about global warming. This may not be conducive to helping students restructure related alternative conceptions. As the few transcript excerpts show, it appears difficult not to include content on ozone layer depletion while teaching about global warming. Perhaps most important is an instructional emphasis on clarifying the causes of and differences between ozone layer depletion and global warming and that neither the ozone "hole" nor consequent increases in UV rays are established causes of global warming. Our findings suggest such clarification should pay particular attention to the following: (a) the differences between incoming (solar) and outgoing (as emitted by earth) radiation; (b) the absorption by greenhouse gases of the latter (but not the former) as *the* mechanism that underlies the greenhouse effect; (c) enhancement of the greenhouse effect due to atmospheric increases in greenhouse gases as *the* cause of global warming; (d) that CFCs, but *not* carbon dioxide, have *two* undesirable effects on our atmosphere (i.e., action as a greenhouse gas and action in the destruction of stratospheric ozone), but that each of these effects cause a *different* environmental problem (enhanced greenhouse effect and ozone layer depletion, respectively); and (e) household aerosol "spray" cans in the United States do not contain CFCs.

These clarifications were incorporated into the STS unit, *Global Atmospheric Change: Enhanced Greenhouse Effect, Ozone Layer Depletion and Ground Level Ozone Pollution* (Rubba, Wiesenmayer, Rye, McLaren, Sillman, Yorks, Yukish, Ditty, Morphew, Bradford, Dorough, and Borza, 1995, http://www.ed.psu.edu/CI/Papers/sts/gac-main.html). The unit actually is a teacher resource that includes the plans for 32 lessons, each lasting for from one to four 40 minute class periods (over 50 days of instruction in total). These are organized into foundations and awareness, investigations, and actions sections (see Table 1).

There are 11 foundations–awareness lessons in which learners examine the nature of science and technology, and characteristic interactions among them that sometimes result in STS issues. Learners build on their understanding about science, technology and society as they develop effective action strategies later in the actions lessons. Significant issues facing humankind are explored, and the STS issue of global atmospheric change (GAC) as it relates to enhanced greenhouse effect, ozone layer depletion,

Table 1. GAC STS Unit Table of Contents

and ground level ozone pollution is identified and expressed as an STS focusing question by learners in the latter lessons of this section.

Learners use the 15 investigations lessons to develop a wide range of inquiry skills for thoroughly exploring STS issues as they apply those skills in investigating GAC. This includes exploring science and social studies concepts and aspects of technology that are foundational to understanding GAC, including, among other topics: the structure, composition and temperature regulation in the atmosphere, the electromagnetic spectrum and energy, ozone layer chemistry and depletion, ground level ozone pollution, reflection and absorption of electromagnetic energy including albedo, the greenhouse effect, human-produced greenhouse gases and the enhanced greenhouse effect, ozone and the greenhouse effect, carbon dioxide detection and analysis, natural sinks for atmospheric carbon dioxide, carbon dioxide and global temperature through time, and the possible impact of lifestyles on global warming. The investigations lessons were designed to present concepts in a specific order within and across these lessons to facilitate conceptual change with regard to misconceptions on enhanced global warming and ozone layer depletion. Information and data from the investigations that involved science and social science methodologies are consolidated by learners in the last investigations lesson to answer the STS focusing question on GAC.

There are six actions lessons. This section is geared toward helping learners develop an understanding of various types of actions that might be taken in support of the answer they formulated to the STS focusing question on GAC. A tentative action plan is composed and the pros and cons associated with each action examined from a number of perspectives. Learners decide which action(s) they are willing to take as individuals or as members of a group. They implement those actions, evaluate the results, and report them to the class.

Each lesson plan includes: overview and outcomes, background notes for the teachers, materials, preparation, instructional procedures, and assessment–portfolio items. Each lesson also includes a concept map, which shows model relationships among concepts the lesson seeks to develop. Also in each lesson, are "background notes" on science, technology, and social aspects of GAC. Teacher-relevant references frequently are noted at the end of the lesson. Five appendices provide resource information for teachers. An overview of concept maps and concept mapping as an instructional tool is among them.

Anecdotal evidence we have from institute teacher-participants is that this unit is effective in helping students clarify the causes of and differences between ozone layer depletion and global warming. There is a need for studies that investigate in what ways students' understandings about global warming change as a result of formal instruction, especially studies that examine the impact of instruction in global climate change on students' conceptions of global warming and ozone layer depletion.

The degree to which learners are able to construct scientific understandings about global warming during instruction may be influenced by, among other things: (a) the tenacity of alternative conceptions held prior to formal instruction about global warming; (b) the instructional model employed by the teacher; (c) alternative conceptions they develop during classroom instruction to help instruction "make sense;" and (d) their level of cognitive development relative to the capacity for formal or abstract thinking (Benson, Wittrock, and Baur, 1993; Driver, Guesne, and Tiberghien, 1985; Eylon and Linn, 1988; Osborne and Freyberg, 1985; Wandersee *et al.*, 1994). According to Piage's theory of cognitive development, middle level students (ages 12–14 years) are making the transition from being concrete operational to formal operational thinkers (Eylon and Linn, 1988; Hassard, 1992).

As previously presented in this chapter, the coverage of the ozone layer "hole" by the media is another potential influence on the construction by learners of alternative conceptions surrounding global warming (Hanif, 1995). Consider the following excerpt from *The New York Times* entitled "Can Capitalism Save the Ozone" (Nasar, 1992):

> But it [eliminating energy subsidies] would also cut emissions of carbon, the substance whose buildup, scientists fear, could eventually destroy the ozone layer, resulting in higher average temperature and harmful climate changes around the world. (p. D2)

This excerpt is perhaps an extreme example of the media's potential to facilitate inappropriate understandings about scientific phenomena. The paper's next edition "corrected" this error ("Corrections," 1992). However, it speaks directly to principal alternative conceptions found to be prevalent amongst students in our study—that carbon dioxide or exhaust causes ozone layer destruction and that ozone layer depletion is a major cause of global warming. Outside the classroom of one of the institute teacher-participants was a poster, intended to inform high school students about the need to use sunscreen, which precipitated the alternative conception that ozone layer depletion was the cause of enhanced global warming. We have observed other similar incidents in TV commercials and magazine ads. CFC-induced ozone layer destruction (and consequent increases in ultraviolet radiation) have received considerable attention from the popular media (Haniff, 1995; Pomerance, 1989).

Global warming is a complex phenomenon: An accurate understanding of the causation of global warming involves abstract concepts about the

electromagnetic spectrum, such as the absorption and re-emission of electromagnetic energy of different wavelengths, the inverse proportionality
that underlies electromagnetic wavelength and energy level, and the relationship of "long-wave" infrared energy to heat and the "greenhouse
effect." The science that underlies global warming and ozone layer depletion is complex and still unfolding (Houghton *et al.*, 1992, 1995; Mackenzie
and Mackenzie, 1995; Mackenzie, 1998; Thiemens, Jackson, Zipf, Erdman,
and Egmond, 1995). Although it is important for science teachers to keep
current on global warming and ozone layer depletion and to reflect major
scientific thinking and consensus in related classroom instruction, we
believe the teachers' efforts to present "all the details" may be counterproductive to students' construction of a clear understanding of global
warming. It is more important for teachers to help students realize the
science and technology that underlies our understanding of complex phenomena and STS issues, such as global warming and its interrelationships
with ozone layer depletion, are somewhat tenuous and may change as scientific studies reveal new data, and explanations are refined (Horner and
Rubba, 1978, 1979).

We recommend classroom instruction on global warming be preceded
by efforts to disclose student understanding of the aforementioned and
involve the use of conceptual change instructional strategies (Smith,
Blakeslee, and Anderson, 1993; Wandersee, *et al.*, 1994) to facilitate the
restructuring of alternative conceptions found to be prevalent among the
students in this study. Accordingly, aspects of the Generative Learning
Model (Osborne and Freyberg, 1985) and concept mapping (Novak
and Gowin, 1984) have considerable application in curriculum design and
instruction on global warming, as well as research (Rye, 1995; Rye and
Rubba, 1998).

Despite the challenges associated with providing global climate change
instruction, the topic is among those global environmental change issues that
is and will continue to be "high profile" (Potts *et al.*, 1996, p. 57). The second
(and most recent) IPCC assessment on climate change contends that, and
for the first time, "[T]he balance of evidence suggests that there is a discernible human influence on global climate" (IPCC, 1995, p. 22), and that
climate change likely will continue. Many (but not all) scientists endorse the
IPCC reports and call for action, observing that: "[T]he further accumulation of greenhouse gases commits the earth irreversibly to further global climatic change and consequent ecological, economic and social disruption"
("Scientists' Statement of Global Climatic Disruption," June 18, 1997).

Instruction on STS issues associated with global climate change can
reveal to students the integrated nature of the science disciplines and an
understanding of related cross-disciplinary concepts (e.g., systems, models,

and change) critical to citizenship scientific literacy (Rutherford, 1989). Such instruction facilitates critical thinking and debate about opposing viewpoints (Roleff, 1997), and provides opportunities for citizenship action. Further, such education can make a crucial contribution to global scientific literacy and from an earth systems perspective—a much needed emphasis according to Mayer (1997): "Science is our attempt to understand our planet. It would be difficult to discern this purpose from most of the world's typical secondary-school science curricula" (p. 102).

IMPLICATIONS FOR THE NEW MILLENNIUM

Although the specific consequences of human activity remain ambiguous, our ability to alter the atmosphere is incontestable. Along with many experts, we believe the gaseous byproducts of modern civilization (e.g., carbon dioxide, CFCs, methane, nitrous oxide, and water vapor) that were dumped in the atmosphere have trapped heat, and so are responsible for the increase in average surface air temperature observed over the past century. While not unknown in intensity over earth's history, this increase has been more rapid than what has occurred in the past 100,000 years, as indicated by natural records. Still, it is possible that this century's temperature increase may be due to natural variation. Admittedly, the supercomputer models used to make predictions have limitations and are based on limited data.

Whether or not there is a discernible human influence on global climate in the form of global warming will continue to be debated by experts and laypeople alike into the next century, even while research provides more data and computer technology provides better modeling capability, but more so as various technical and social resolution paths and coping strategies are proposed and implemented. Global warming potentially may be the most serious problem humankind has ever faced. Our ability to deal responsibly with this issue will test the science, technology and social structures of the global community. Global warming will be the paramount STS issue beyond 2000. It will be the STS issue that shapes STS education as STS enters the next millennium.

REFERENCES

Abimbola, I. (1988). The problem of terminology in the study of student conceptions in science. *Science Education* 72:175–184.

Activities for the Changing Earth System (1993). Ohio State University Research Foundation, Columbus, OH.

American Association for the Advancement of Science (1990). *Science for All Americans*, Oxford University Press, New York.

American Association for the Advancement of Science (1993). *Benchmarks for Science Literacy*, Oxford University Press, New York.

Ashmore, M. R., and Bell, J. N. B. (1991). The role of ozone in global change. *Annals of Botany* 67:39–48.

Austin, J., Butchart, N., and Shine, K. (1992). Possibility of an Arctic ozone hole in a doubled-CO_2 climate: *Nature* 360:221–225.

Benson, D. L., Wittrock, M. C., and Baur, M. E. (1993). Students' preconceptions of the nature of gases. *Journal of Research in Science Teaching* 30:587–597.

Boyes, E., and Stanisstreet, M. (1993). The greenhouse effect: Children's perceptions of causes, consequences, and cures: *International Journal of Science Education* 15:531–552.

Christidou, V. (1994). An exploration of children's models and their use of cognitive strategies in regard to the greenhouse effect and the ozone layer depletion. Paper presented at the 2nd European Summer School-Research in Science Education. Leptocaria. Greece.

Christidou, V., and Koulaidis, V. (1996). Children's models of the ozone layer and ozone depletion. *Research in Science Education* 26:421–436.

Corrections (1992). *The New York Times*, February 8, 1992, p. A3.

Denman, K., Hofmann, E., and Marchant, H. (1996). Marine biotic responses to environmental change and feedbacks to climate. In Houghton, J., Filho, L., Callander, B., Harris, N., Kattenberg, A., and Maskell, K. (eds.), *Climate Change 1995*, The Science of Climate Change, Press Syndicate of the University of Cambridge, New York.

Dorough, D., Rubba, P., and Rye, J. (1995). Fifth and sixth grade students' explanations of global warming. Paper presented at the 1995 Annual National Association for Research in Science Teaching Meeting, San Francisco, CA, April, 1995.

Driver, R., Guesne, E., and Tiberghien, A. (1985). *Children's Ideas in Science*, Open University Press, Philadelphia, PA.

Ennis, C., and Marcus, N. (1994). *Biological Consequences of Global Climate Change*, University Corporation for Atmospheric Research, Boulder, CO.

Eylon, B., and Linn, M. C. (1988). Learning and instruction: An examination of four research perspectives in science education. *Review of Educational Research* 58:251–301.

Francis, C., Boyes, E., Qualter, A., and Stanisstreet, M. (1993). Ideas of elementary students about reducing the "greenhouse effect." *Science Education* 77:375–392.

Fishman, J. (1991). The global consequences of increasing tropospheric ozone concentrations. *Chemosphere* 22:685–695.

Hanif, M. (1995). Understanding ozone. *The Science Teacher* 62:20–23.

Hassard, J. (1992). *Minds on Science: Middle and Secondary School Methods*, Harper Collins, New York.

Heinze-Fry, J., and Novak, J. (1990). Concept mapping brings long-term movement toward meaningful learning. *Science Education* 74:461–472.

Hocking, C., Sneider, C., Erickson, J., and Golden, R. (1990). *Global Warming and the Greenhouse Effect* (A curriculum guide for grades 7–10). The Regents of the University of California, Berkeley.

Horner, J., and Rubba, P. (1978). The myth of absolute truth. *The Science Teacher* 45(1):29–30.

Horner, J., and Rubba, P. (1979). The laws-are-mature-theories fable. *The Science Teacher* 46(2):31.

Houghton, J., Jenkins, G., and Ephraums, J. (eds.). *Climate Change: The IPCC Scientific Assessment*, Cambridge University Press, Cambridge, MA.

Houghton, J., Callander, B., and Varney, S. (eds.) (1992). *Climate Change 1992. The Supplementary Report to the IPCC Scientific Assessment*, Press Syndicate of the University of Cambridge, New York. pp. 1–22.

Houghton, J., Filho, L., Callander, B., Harris, N., Kattenberg, A., and Maskell, K. (eds.) (1996). *Climate Change 1995. The Science of Climate Change*, Press Syndicate of the University of Cambridge, New York.

Houghton, J., Filho, M., Bruce, J., Lee, H., Callander, B., Haites, E., Harris, N., and Maskell, K. (1994). *Climate Change 1994: Radiative Forcing of Climate Change and an Evaluation of the IPCC 1992 Emissions Scenarios*, Cambridge University Press, Cambridge, UK.

Houghton, J. T., Callander, B. A., and Varney, S. K. (eds.) (1992). *Climate Change 1992. The Supplementary Report to the IPCC Scientific Assessment*, Press Syndicate of the University of Cambridge, New York. pp. 1–22.

IPCC (1995). *Climate Change 1995: IPCC Second Assessment Report*, World Meterological Organization, Geneva, Switzerland.

Jonassen, D. H., Beissner, K., and Yacci, M. (1993). *Structural Knowledge*. Lawrence Erlbaum Associates, Hillsdale, NJ.

Kattenberg, A., Giorgi, H., Grassl, H., Meehl, G., Mitchell, J., Stouffer, R., Tokioka, T., Weaver, A., and Wigley, T. (1996). Climate models—Projections of future climate. In Houghton, J., Filho, L., Callander, B., Harris, N., Kattenberg, A., and Maskell, K. (eds.), *Climate Change 1995. The Science of Climate Change*, Press Syndicate of the University of Cambridge, New York. pp. 285–357.

Kerr, R. (1997). The right climate for assessment. *Science* 277:1916–1918.

Kotlyakov, V. (1996). Climatic change and the future of the human environment. *International Social Science Journal* 150:511–523.

Koulaidis, V., and Christidou, I. (1993). Children's misconceptions and cognitive strategies regarding the understanding of the ozone layer depletion. Abstract of paper presented at the Third International Seminar on Misconceptions and Educational Strategies in Science and Mathematics, Ithaca, NY, August, 1993.

Kurtis Productions. (1993). *Ozone: The hole story*. (Teaching unit and videocassette.) Kurtis Productions, St. Petersburg, FL.

Lacis, A., Wuebbles, D., and Logan, J. (1990). Radiative forcing of climate by changes in the vertical distribution of ozone. *Journal of Geophysical Research* 95(D7):9971–9981.

Lomask, M., Baron, J., and Grieg, J. (1993). Assessing conceptual understanding in science through the use of two- and three-dimensional maps. A paper presented at the Third International Seminar on Misconceptions and Educational Strategies in Science and Mathematics, Ithaca, NY, August, 1993.

Mayer, V. (1997). Global science literacy: An earth system view. *Journal of Research in Science Teaching* 34:101–105.

Mackenzie, F., and Mackenzie, J. (1995). *Our Changing planet: An Introduction to Earth System Science and Global Environmental Change*. Prentice Hall, Englewood Cliffs, NJ.

Mackenzie, F. (1998). *Our Changing Planet—An Introduction to Earth System Science and Global Environmental Change* (2nd ed.). Prentice-Hall, Upper Saddle River, NJ.

Metzger, E. (1996). The STRATegy column for precollege science teachers—Charting global climate change. *Journal of Geoscience Education* 44:327–333.

Mitchell, J. F. (1989). The "greenhouse" effect and climate change. *Reviews of Geophysics* 27:115–139.

Monastersky, R. (1992). Up, up and fading away. *Science World* 48:6–9.

Nasar, S. (1992). Can capitalism save the ozone? *The New York Times*, February 7, 1992, p. D2.

National Academy of Sciences (1992). *Policy Implications of Greenhouse Warming*. National Academy Press, Washington, D.C.

National Research Council (1996). *National Science Education Standards*. Washington, National Academy Press.

Novak J., and Gowin, D. (1984). *Learning How to Learn*. Cambridge University Press, New York.

O'Meara, M. (1997). The risks of disrupting climate. *World Watch* 10(6):10–24.

Osborne, R., and Freyberg, P. (1985). *Learning in Science*. Heinemann Education, Auckland.

Patton, M. Q. (1987). *How to Use Qualitative Methods in Evaluation*. Sage Publications, Beverly Hills.

Patton, M. Q. (1990). *Qualitative Evaluation and Research Methods*. Sage Publications, London.

Plunkett, S., and Skamp, K. (1994). The ozone layer and hole: Children's misconceptions. A paper presented at the Australian Science Education Research Association Conference, Hobart, Tasmania, July, 1994.

Potts, A., Stanisstreet, M., and Boyes, E. (1996). Children's ideas about the ozone layer and opportunities for physics teaching. *School Science Review* 78:57–62.

Pomerance, R. (1989). The dangers from climate warming: a public awakening. In Abrahamson (ed.), *The Challenge of Global Warming*, Island Press, Covelo, CA, pp. 259–269.

Roberts, D. (1992). *Statistical Analysis for the Social Sciences*. Kendall-Hunt Publishers, Dubuque.

Roleff, T. (1997). *Global Warming*. Greenhaven Press, San Diego.

Rubba, P., and Wiesenmayer, R. (1988). Goals and competencies for precollege STS education: Recommendations based upon recent literature in environmental education. *The Journal of Environmental Education* 19:38–44.

Rubba, P., Rye, J., and Wiesenmayer, R. (1994). Integrating STS units into middle/junior high school science: Insights gained from teacher-developers. A paper presented at the Association for the Education of Teachers of Science Annual Meeting, El Paso, TX, January, 1994.

Rubba, P., and Wiesenmayer, R. (1985). A goal structure for precollege STS education: A proposal based upon recent literature in environmental education. *Bulletin of Science, Technology and Society* 5:573–580.

Rubba, P., and Wiesenmayer, R. (1993). Increased actions by students. In Yager, R. E. (ed.), *The Science, Technology, Society Movement*, National Science Teachers Association, Washington, D.C., pp. 169–175.

Rubba, P., and Wiesenmayer, R. (1997). Goals and competencies for precollege STS education: Recommendations based upon recent literature in environmental education. In *Essential Research in Environmental Education*, Stipes Publishing L.L.C., Champaign, IL, pp. 327–336. (Reprinted from: *Journal of Environmental Education* 19(4):38–44, 1988 with permission.)

Rubba, P., Wiesenmayer, R., Rye, J., and Ditty, T. (1996). The leadership institute in STS education: A collaborative teacher enhancement, curriculum development and research project of Penn State University and West Virginia University with rural middle/junior high school science teachers. *Journal of Science Teacher Education* 7:1–10.

Rubba, P., Wiesenmayer, R., Rye, J., McLaren, M., Sillman, K., Yorks, K., Yukish, D., Ditty, T., Morphew, V., Bradford, C., Dorough, D., and Borza, K. (1995). *Global Atmospheric Change: Enhanced Greenhouse Effect, Ozone Layer Depletion and Ground Level Ozone Pollution*, The Pennsylvania State University, University Park, PA.

Rye, J. (1995). An Exploratory Study of the Concept Map as an Interview Tool to Facilitate Externalization of Conceptual Understandings Associated with Global Atmospheric Change by Eighth Grade Physical Science Students. (Doctoral dissertation, The Pennsylvania State University, University Park).

Rye, J. (1998). Understanding the role of chlorofluorocarbons in global atmospheric change. *Journal of Geoscience Education* 46:488–493.

Rye, J., Rubba, P., and Wiesenmayer, R. (1997). An investigation of middle school students' alternative conceptions of global warming. *International Journal of Science Education* 19:527–551.

Rye, J., and Rubba, P. (1998). An exploratory study of the concept map as a tool to facilitate the externalization of students' understandings about global atmospheric change in the interview setting. *Journal of Research in Science Teaching* 35:521–546.

Rutherford, J. (1989). *Science for All Americans*, Oxford University Press, New York.

Sanhueza, E., and Zhou, X. (1996). Radiative forcing of climate change (section 2.2). In Houghton, J. *et al.* (ed.), *Climate Change 1995*, The Science of Climate Change, Press Syndicate of the University of Cambridge, New York.

Schmid, R. (1998). 1997 was hottest year on record. The wire: the global gamble, http://wire.ap.org/APnews/center?FRONTID=PACJAGE&EXTRA-global warming.

Schneider, S. H. (1989). The changing climate. *Scientific American* 261:70–79.

Science for Understanding Tomorrow's World: Global Change (1994). Paris, France: International Council of Scientific Unions.

Scientists' statement on global climatic disruption (1997). Ozone Action, Washington, D.C. http://www.ozone.org/stateii.html.

Smith, E. L., Blakeslee, T. D., and Anderson, C. W. (1993). Teaching strategies associated with conceptual change learning in science. *Journal of Research in Science Teaching* 30:111–126.

Solomon, S., and Srinivasan, J. (1996). Radiative forcing of climate change (section 2.4). In Houghton, J., Filho, L., Callander, B., Harris, N., Kattenberg, A., and Maskell, K. (eds.), *Climate Change 1995*, The Science of Climate Change, Press Syndicate of the University of Cambridge, New York.

Thiemens, M., Jackson, T., Zipf, E., Erdman, P., and Egmond, C. (1995). Carbon dioxide and oxygen isotope anomalies in the mesosphere and stratosphere. *Science* 270:969–972.

U.N.: Ozone thinner than ever, 1997 June 6: The Dominion Post, p. 5a. (Quoting Scientific American Library, Ultimate Visual Dictionary, AP Research).

United States Environmental Protection Agency, 1995 (March), Stratospheric ozone depletion: Washington, D.C., Office of Research and Development, 24 p. (EPA/640/K-95/004).

University Corporation for Atmospheric Research (1992). *Our ozone shield*. (Reports to the Nation on Our Changing Planet, No. 2). Boulder: Office of Interdisciplinary Earth Studies.

Waks, L., and Prakash, M. (1985). STS education and its three step-sisters. *Bulletin of Science, Technology and Society* 5:105–116.

Wandersee, J., Mintzes, J., and Novak, J. (1994). Research on alternative conceptions in science. In Gabel, D. (eds.), *Handbook of Research on Science Teaching and Learning*, MacMillan Publishing Company, New York pp. 177–210.

Warrick, J. (1998, January 9). Scientists see weather trend as powerful proof of global warming. *The Washington Post*, A8.

Warrick, R., Provost, C., Meier, M., Oerlemans, J., and Woodworth, P. (1996). Changes in sea level. In Houghton, J., Filho, L., Callander, B., Harris, N., Kattenberg, A., and Maskell, K. (eds.), *Climate Change 1995*. The Science of Climate Change, Press Syndicate of the University of Cambridge, New York pp. 359–405.

West, L. H., and Pines, A. L. (eds.) (1985). *Cognitive Structure and Conceptual Change*, Academic Press, Inc., Orlando, FL.

Wise, K., and Okey, J. R. (1983). A meta-analysis of the effects of various science teaching strategies on achievement. *Journal of Research in Science Teaching* 20:419–435.

World Meteorological Organization (1995). *Scientific Assessment of Ozone Depletion: 1994 Executive Summary*, World Meteorological Organization (Report No. 37), Geneva, Switzerland.

Worrest, R. C., Smythe, K. D., and Tait, A. M. (1989). *Linkages Between Climate Change and Stratospheric Ozone Depletion.* (Report No. EPA/600/D-89/127) U.S. Environmental Protection Agency, Washington, D.C.

CHAPTER 9

STS Education for Knowledge Professionals

J. Scott Hauger

This chapter considers the value of an education in science, technology, and society studies (STS) for managers or analysts in knowledge-based industries, government, and other professions who are concerned with managing the course of technological change. By technological change, I mean intentional social innovation in the form of new tools and new ways of using them. The term "knowledge professional" is appropriate to describe these workers, because the kinds of social innovation they analyze or manage intimately involves the creation, synthesis, dissemination and application of new, science-based knowledge. (I wavered between the term "knowledge professional" and "innovation professional" to describe the workers I am discussing here. Knowledge professional seemed too broad, perhaps too closely related to the common term "knowledge worker." It might be taken to include data entry specialists or anyone else whose occupational use of "knowledge" is passive or routine. Innovation professional seemed too narrowly focused on those who work specifically to adopt a new technology, thus omitting, for example, research program managers,

J. Scott Hauger, Research Competitiveness Program, American Association for the Advancement of Science, Washington, D.C. 20005.

Science, Technology, and Society: A Sourcebook on Research and Practice, edited by Kumar and Chubin, Kluwer Academic / Plenum Publishers, New York, 2000.

research administrators and science policy analysts, all of whom I would intend to include.)

THE EMERGENCE OF KNOWLEDGE PROFESSIONALS

I first became aware of the emerging group of knowledge professionals, about twenty-five years ago, not through study, but on the job. By this time, a small industry had developed around the Washington beltway to provide technology assessment, systems development, and systems integration services, primarily to the Department of Defense. The Defense Department had adopted a strategy for global competition with the Soviet Union that relied on technological innovation to achieve a strategic superiority over an opponent which was thought to have an advantage in terms of numbers, material, and initiative. The military services and the various component agencies of the Department of Defense sought to stimulate, promote, and adopt new technologies in many areas—from weapons to support services, from health care to logistics—that would provide a competitive advantage. More than most institutions, the Defense Department was aware that technological innovation entailed social innovation, from the structure of military organizations, to the training of personnel, to planning for the interoperability of adjacent technologies.

The industry that emerged to meet the demand for strategically planned innovation was eclectic in the disciplinary background of its practitioners. The problems it was called upon to address were complex in scope. Their solution required both specialists in many disciplines and team leaders who could not only provide a conceptual framework for problem solving but also synthesize knowledge across disciplines. The new industry may have centered on engineers, but it included economists, physicists, modelers, managers, even historians—anyone who by some combination of their education and prior professional experience could embrace both the technical and social, usually organizational, elements of innovation.

By the 1970s, the talent assembled in "professional services firms," more informally known as "think tanks," or derisively as "beltway bandits," was being applied to the solution of problems well beyond the defense sector. Great Society programs led to the pursuit of innovation in education, transportation, and disability fields. The Arab oil embargo and the creation of the Department of Energy led to programs in conservation and alternative energy. Certain bureaus and divisions in most federal government agencies were charged with or created for the management of innovation. Congress created the Office of Technology Assessment (OTA) to be an inhouse think tank concerned with innovation. OTA was staffed with knowledge professionals. Meanwhile, the firms which provided new products and processes in response to such government-sponsored innovation were also hiring or creating new knowledge professionals who could respond to the opportunities by writing winning proposals and managing the resulting projects.

During the last 15 years, two important factors have broadened the range of the knowledge professional. At the same time, they contributed to a convergence of interests among the government-oriented professionals, with whom I worked in the 1970s and 1980s, and private sector managers concerned with innovation as a business strategy. One is the development of the desktop computer, beginning about 1980, and all the subsequent innovations which are currently manifest in the availability and use of spreadsheets, flow charts, e-mail, databases and information available on the worldwide web. These new tools enabled knowledge professions to be practiced at any desk in any office or home, in collaboration with other professionals in any other location, without requiring a large support staff to gather data, make computer runs, type and print reports, produce slides, and disseminate findings. Every firm, every division of a firm or agency now has the ability to support or hire a knowledge professional, and in an interesting case of reflexivity, many firms and agencies have hired knowledge professionals to help them adopt and manage the new computer-based technologies.

The other important factor in the spread of knowledge professionals is the end of the Cold War and the relative decline of the federal government as the motive force for competitive innovation, a role which has been shifting toward the private sector. A strategy of economic competition through technological innovation has long characterized many American businesses which have sought to implement new, science-based production technologies to overcome the advantage of inexpensive labor in other countries. The rise of Japanese competitors, among others, in the sale of new technology products has provided new incentives to institutionalize business strategies which depend upon continuous innovation. Daily advances in a number of science-based technologies from microprocessors to materials to tailored genes make such a strategy feasible at a time when the end of the Cold War is supporting a return to the proposition that the business of America is business. This trend is reinforced by the growth of the information technologies sector which supports a general speed up in the product innovation cycle.

Strategic planning for innovation has become a major concern for the private sector, both as producers and consumers of innovation. Firms believe they cannot remain competitive merely by responding to the market. Like the Defense Department in the Cold War, business firms in many sectors seek to lead the market into a new technological context as a strategy to remain competitive on a global level. Moreover, dramatic failures in the innovation of otherwise successful technologies, from beta-format video recordings to nuclear fission power plants, have provided industry with clear reasons to believe that the success of innovation strategies depends on much more than engineering knowledge. Disciplinary experts or practitioners who rely on traditional, specialized practices are not sufficient to ensure success in the current business environment. Business firms are creating or hiring their own knowledge professionals, sometimes importing them from the institutions created during the Cold War. The employment of knowledge professionals also appears to be increasing within state government agencies and institutions, such as land grant universities, as the states assume such expanded roles as promoting economic development through innovation and cost-sharing of federally-sponsored research.

WHO ARE KNOWLEDGE PROFESSIONALS?

My personal experiences and observations of the emergence of knowledge professional are in no way privileged or unique. Social scientists and commentators focusing on business and labor have been writing about the changing structure of the workforce for many years. In *The Work of Nations*, Robert Reich (1991) devotes six chapters to "the rise of the symbolic analyst." According to Reich, this job category, includes

> . . . all the problem-solving, problem-identifying, and strategic-brokering activities. . . . of many people who call themselves research scientists, design engineers, software engineers, civil engineers, biotechnology engineers, sound engineers, public relations executives, investment bankers, lawyers, real estate developers, and even a few creative accountants. Also included is much of the work done by management consultants, financial consultants, tax consultants, agricultural consultants, armaments consultants, architectural consultants, management information specialists, organization development specialists, strategic planners, corporate headquarters and systems analysts. Also: advertising executives and marketing strategists, art directors, architects, cinematographers, film editors, production designers, publishers, writers and editors, journalists, musicians, television and film producers, and even university professors. (177)

Reich describes symbolic analysts as those who "solve, identify, and broker problems by manipulating symbols. . . . The tools may be mathematical algorithms, legal arguments, financial gimmicks, scientific principles, psychological insights about how to persuade or amuse, systems of induction or deduction, or any other set of techniques for doing conceptual puzzles" (178). He states that in the 1980s, about 20 percent of American workers fell under this category up from about eight percent in the 1950s (179). "As the value placed on new designs and concepts continues to grow relative to the value placed on standard products," Reich claims, "the demand for symbolic analysts will continue to surge" (225). In separate chapters, Reich discusses the education of symbolic analysts, and asks, ". . . how the considerable skills and insights of symbolic analysts can be harnessed for the public good" (185). His consideration of the education of symbolic analysts is largely descriptive rather than normative. He notes that symbolic analysts are being taught four basic skills: abstraction of a problem, system thinking, experimentation, and collaboration with other symbolic analysts (225–233). Reich also notes the importance of on-the-job training of symbolic analysts working in common problem areas, often located in sub-communities which he calls "symbolic analysis zones" (235).

It is clear that Reich's job category of symbolic analyst is a much larger one than that of knowledge professional, as considered here. Based upon his examples and descriptions, however, it seems reasonable to identify knowledge professionals as a subset of symbolic analysts. Specifically, knowledge professionals are those who identify, solve and broker problems of science-based technological innovation. In Reich's terms, the Washington beltway in the 1970s and 1980s, comprised a symbolic analysis zone for knowledge professionals, providing innovation services to federal government agencies, while providing on-the-job training for careers in identifying, solving, and brokering problems of government-sponsored technological innovation.

In 1959, Peter Drucker coined the phrase, "knowledge worker" to describe a "newly emerging dominant group," displacing the importance of industrial workers in the post-industrial age (1994, 6). Like Reich's symbolic analysts, Drucker's knowledge workers comprise a larger category than the knowledge professionals considered here. Knowledge workers, whom Drucker characterizes but leaves undefined, employ knowledge as a tool to solve a problem at hand. They function as employees but control their own tools of production. The value of their knowledge to the market depends upon the task at hand. They gain access to jobs and social position only through formal education, a fact that, according to Drucker, has important social implications: "Education will become the center of the knowledge society, and the school its key institution," although ". . . more and more knowledge, and especially advanced knowledge, will be acquired well past the age of formal schooling and increasingly, perhaps, through educational processes that do not center on the traditional school" (9).

Drucker holds to the reductionist notion that knowledge must be specialized. He believes knowledge workers are, by definition, specialists. He says the term "generalist" will have no professional meaning other than ". . . people who have learned to acquire additional specialties rapidly in order to move from one kind of job to another—for example, from market research into management, or from nursing into hospital administration. But 'generalists' in the sense in which we used to talk of them are coming to be seen as dilettantes rather than educated people" (10).

This characterization of knowledge workers might seem antithetical to an STS education which is grounded in the liberal arts and seeks to produce a certain kind of generalist. We need not accept Drucker's assertion at face value, however. As anticipated by his examples of specialists in market research and nursing who acquire additional knowledge to assume managerial roles, Drucker goes on to describe a new kind of generalist: the manager, who must orchestrate the performance of knowledge specialists to make their knowledges productive. STS scholars might well understand Drucker's manager as a *meta-knowledge worker*, one who uses a knowledge of knowledge work as the basis of his or her professional practice.

Drucker's argument runs something like this: Because of the need for specialization in discipline or practice, knowledge workers must work in multidisciplinary teams to solve societal problems and perform tasks which cut across knowledge disciplines: "With knowledge work growing increasingly effective as it is increasingly specialized, teams become the work unit rather than the individual himself" (11). For Drucker, this situation enhances and makes visible the importance of management, which can be understood as the practice of bringing people possessing different knowledge together for a joint performance (13). He concludes "Management is still taught in most business schools as a bundle of techniques, such as budgeting and personal relations. To be sure, management, like any other work has its own tools and techniques. But just as the essence of medicine is not urinalysis (important though that is), the essence of management is not techniques and procedures. The essence of management is to make knowledges productive. Management in other words is a social function. And in its practice, management is truly a liberal art" (13–14).

Drucker's phrase, "managers who make knowledges productive," is a good description of knowledge professionals, as I have observed them. Thus, knowledge professionals are a subset of knowledge workers, as discussed by Drucker, and of symbolic analysts as discussed by Reich. They are professionals whose social function is to identify, solve, and broker problems of science-based technological innovation, by leading or otherwise combining the knowledges of multidisciplinary team members to make knowledge productive with respect to the problem at hand.

It is a bit difficult to further specify or define this group. Indeed, they lack many of the characteristics that typically define a group of professionals. They come to their area of common interest from a variety of directions. They do not as yet identify themselves with a common set of social institutions. They have undertaken their college educations in a variety of disciplines and fields. They work in many different settings. Their job titles vary. They do not belong to a single set of professional organizations. Society has yet to develop a common label for them. Nonetheless, I meet them every day, both in my job at the American Association for the Advancement of Science (AAAS) and in the classes I teach at Virginia Tech's Northern Virginia Center (NVC). They may also be found in a broad, but overlapping variety of organizations, conferences and meetings devoted to aspects of innovation. They tend to be mobile within a common set of firms and institutional employers. The common characteristic of these knowledge professionals is that their jobs and careers are largely concerned with the integration of new, science and technology-based knowledge into the workplace.[1]

AT WORK WITH KNOWLEDGE PROFESSIONALS

If it were not for my work with AAAS, I would believe the knowledge professional is a particular denizen of the Washington symbolic analysis zone. There are, in Washington, many professionals concerned with governmental aspects of scientific research and technological change. Because Washington is the national center for science and technology policy, it is also the center for science and technology policy-related professions. The Washington-area positions are labeled as science policy jobs, but they have much in common with jobs around the country which do not share the label.

My position as director of AAAS's Research Competitiveness Program is explicitly labeled a science policy job. It is, however, a job that is concerned with activities far from the halls of the federal government. The program works with states that are least engaged in federally—sponsored scientific research and technology development, to enhance the states' "research competitiveness." The program, sponsored by the National Science Foundation's Experimental Program to Stimulate Competitive Research (EPSCoR) has taken me such places as Jackson, Mississippi, Burlington, Vermont, and Laramie, Wyoming. At each location, I found a variety of professionals concerned with issues and problems of science-based technological and social change, although the label of science and technology policy is rarely applied. They include, for example, university administrators concerned with issues of industry-sponsored research, public interest group staff members working on environmental problems, businessmen dealing with technology issues related to the global marketplace, local government officials concerned about the creation of high-paying jobs, and educators concerned with issues such as distance education. Like their colleagues within the Washington beltway, these knowledge professionals

---

[1] I use the term *workplace* broadly to include any place people work, be it the office, military headquarters, classroom, home, or laboratory.

are working daily to integrate new, science and technology-based knowledge into the workplace. Although these integrative jobs are sometimes an additional task for a teacher, manager, or professional, because of their increasing scope and importance, they often become full-time positions.

The professional positions I am describing may be distinguished from others which are characterized by specialization, either by discipline or by application. Many new knowledge professionals were trained, or started their careers, as specialists, but forces in the social environment caused them to pursue careers outside the boundaries of their disciplinary education—careers which may not even have existed when they received their college degrees or entered the job market. By pursuing the best available opportunities through a series of generally progressive job changes, a growing number of professionals—each following his or her own path—came to establish careers wherein the relationships between successive jobs was defined not by a body of technical knowledge to be mastered by the individual, but by a process of managing, analyzing, and planning for the adoption of new, science-based, technological knowledge into the workplace.

From this perspective, many of the science and technology policy jobs found in Washington can be seen to comprise part of a broader domain of professional positions concerned with technological and therefore social innovation. Such jobs are labeled "policy" jobs when they seek to serve or inform government decision making. They may bear other labels, such as "strategic planning," when they are in the service of business or education. They share a concern for planning or managing the course of science-based technological innovation.

The emergence of a knowledge profession is perhaps more readily discerned in careers than in individual jobs. For example, my own career over the last 20 years, has included employment as a chemist, a military officer, a technology analyst in a think tank, a small business owner, a consultant, an academic research manager, a university professor, and, most recently, by a nonprofit organization. During that time, I was involved in innovation projects in areas including cruise missile defense, large-scale simulation modeling, solar thermal power plants, technologies for persons with disabilities, design for the environment, and bio-based materials, among others. From a traditional perspective such a seeming hodge podge may not look like a career at all. There is no common theme of employing institution, scientific discipline, or technology application area. But there has been a continuity of practice, and my responsibilities have been continuously progressive around a theme of managing the course of research, development and innovation. Similar patterns may be found in the careers of a number of my acquaintances who are in the second half of their professional lives. They have built careers which cut across industry, academia, and government, in a series of progressive jobs which deal with issues and problems of innovation. Responding individually to the common environment of a workplace which has rewarded science-based technological innovation, a significant number of individuals have built careers which might be said to represent a new paradigm for knowledge professionals. Judging from the younger professionals I have met on the job and in the classroom, this has been a growing trend over the last 20 years, extending to more junior levels in organizations, and becoming visible earlier in a professional career.

KNOWLEDGE PROFESSIONALS AND STS

The knowledge professional was created by the workplace and the job market, not by our educational institutions which are, I think, inherently conservative. They will react to changes in the workplace over time, but they have no incentive to lead the way, and harbor insufficient knowledge of nonacademic professions to provide the leading edge of change. The emerging academic field of science, technology and society studies did not set out to educate knowledge professionals. Nor, for the most part, are knowledge professionals the product of an STS education. Nonetheless, STS provides a potential academic base for knowledge professionals who wish to better understand their practices.

Steve Fuller (1997) characterized the STS community as consisting of a high church and a low church:

> In High Church terms, "STS" means "Science & Technology Studies," an emerging academic discipline that uses the methods of the humanities and the social sciences to study mainly the natural sciences but increasingly technology. In Low Church terms, "STS" means "Science, Technology & Society," a nascent social movement that has been historically promoted by science and engineering teachers concerned with the social implications of mainly technology but increasingly science. There is probably a broad political consensus between the High and Low Churches regarding a generally critical attitude toward the role of science and technology in society today. However the High Church stresses the need for more research to understand the complexities of that role, whereas the Low Church wishes to reduce some of those complexities by reorienting science and engineering education. Consequently, the two Churches of STS inhabit rather different professional societies and represent themselves in rather different ways, though often drawing from many of the same intellectual traditions. (no page number)

Fuller, of course, is writing from the academy, rather than the work place, so he misses a Broad Church perspective which would embrace the concerns of the working congregation as well as those of the academic priesthood. Like the Low Church academic reformer, knowledge professionals are concerned with the social implications of ". . . mainly technology, but increasingly science." The social groups they are concerned with are more likely to be offices, firms, agencies, or professional practices than nations or civilizations. But academic STS has also had an interest in microlevel organizations. The social implications of technologies that knowledge professionals would like to understand are often focused on issues of the marketplace and on relations between people in the workplace who employ a new technology, including changes in the way they interact with their neighbors. These issues also fall within the boundaries of scholarly research in STS. Although most knowledge professionals may never participate in the High Church ritual of scholarly research, they are eager for research results that can inform their practices. They comprise potential consumers for STS research products in a way analogous to engineers' consumption of the results of research in the natural sciences. Moreover, some knowledge professionals are both competent and interested in participating in research, and can bring to that enterprise a knowledge of and access to some excellent research sites.

In short, knowledge professionals whose daily practices are concerned with managing the course of science-based technological innovation have a broad interest in the same issues as high and low church academic STS programs. If "critical" be understood to mean "exercising or involving careful judgment or judicious evaluation," rather than "inclined to criticize severely and unfavorably" (Webster, 1985, p. 307), then these potential students typically share ". . . a generally critical attitude toward the role of science and technology in society today." They are paid to exercise their critical skills and judgment on a daily basis.

Most knowledge professionals have not yet heard of academic STS, or are only vaguely aware of the field. I base this assertion on the experience of hundreds of conversations with potential students who inquire about the Virginia Tech graduate program at the Northern Virginia Center. Yet, these inquirers quickly grasp the nature of the program when it is explained, and some fraction of them find it more interesting than competing opportunities to earn a master's degree in such fields as business administration, information technology, engineering, engineering administration, or public affairs and policy administration, even though the STS diploma is a lesser known degree without a clearly associated career path.

Lacking their own common academic base, knowledge professionals may be attracted to the field as an alma mater. This happened to me in mid-career, when I learned, 15 years out of the university, of STS as a field of study. It is happening to my students at a much earlier point in their careers. All of these master's-level students have a bachelor's degree in a field other than STS. (About 50 students have taken courses in the first five semesters of the program.) Between a third and a half have another master's degree. Four people with doctorates have enrolled in these STS courses. Nearly all the students are working professionals—in fields as diverse as nursing, systems engineering, the military, librarianship, telecommunications, project management, and technology assessment, as well as a few students in civil service positions with a science and technology policy dimension. How well can the field of science, technology and society studies provide what these practitioners need, and how will the experience of meeting such a need influence the direction of STS?

At the M.S. level, I will argue, STS can meet the needs of knowledge professionals very well. At the B.S. level there is less opportunity. There are few entry-level positions for knowledge professionals. In Reich's terms, the brokerage of problems in science-based technological innovation requires more experience and a higher level of education than a fresh bachelor's degree can provide. Drucker identifies knowledge professionals with managers, which indicates that we should not expect to find many knowledge professionals in entry-level positions. Combining Reich and Drucker's ideas, we might state that in the contemporary workplace, most entry—level symbolic analysts work to solve problems across one field of knowledge. Later in their careers, symbolic analysts, who are successful at solving problems of a more limited scope and who show a talent for leading teams of specialists, are promoted into positions that require them to work across several fields of knowledge. Students who wish to follow a career of knowledge professional, therefore, would probably be well advised to earn a bachelor's degree in a more specialized field of interest before pursuing graduate education in STS.

This does not mean there is no role for STS in the education of the knowledge professional at the bachelor's level. There is, after all, a difference between a student who pursues a degree, for example in chemistry, to prepare for a career as a chemist and a student who seeks the same degree as a step toward a career in managing technological change. It is debatable whether or not STS courses will make a student a better chemist. But it seems certain to me that STS courses will help a chemist be better at identifying, solving, and brokering problems that include chemistry as one of several knowledges needed to address the problem. Undergraduates who are sensitive to the nature of the workplace as described by Drucker and Reich, and who see their career path as that of a knowledge professional rather than that of a chemist, would be well advised to take elective courses or to construct a minor in STS or its component fields.

There is another good reason for knowledge professionals to pursue an undergraduate major and even an advanced degree in a specialized field of knowledge: credibility. If, together with Steve Fuller, we are to understand STS as a critical field, then we must have sufficient mastery of the disciplines and practices we critique to ensure credibility with ourselves, our employers, and those whose work we wish to analyze and influence. In academic STS, the principal audience for our work is found within the boundary of our own knowledge community. Other STS scholars, as a community of peers, are responsible for evaluating our performance. The evaluation of our work by the scientists and engineers whom we study is secondary at best. Their opinions, especially negative ones, may even be suspect. In the nonacademic workplace, however, other knowledge communities must benefit from our work and find value in it. Credibility is needed for access to these communities, and also for effectiveness in negotiating solutions to problems which they can accept.

The nonacademic workplace imposes different and, I think, difficult standards of successful critical analysis. It favors pragmatic incrementalism and compromise in comparison to a more paradigmatic form of STS. It makes STS more a technology, where new knowledge is a matter of design for the solution of problems presented by the society at large, and less a science, where both problems and solutions are validated by the community of peers. Just as in engineering and architecture, the work of STS-trained knowledge professionals will be judged not only by the professional community but also by the communities they seek to serve. Those communities are more likely to trust a critical approach to science and technology when the critic is well-grounded and credentialed in both critical analysis and in that which is being analyzed. STS-based knowledge professionals can benefit from an undergraduate education in one of the scientific or technological disciplines.

Some STS scholars may object to the compromises an STS-educated knowledge professional must make in order to succeed as a nonacademic practitioner. They may view scientific knowledge and the knowledge of the STS community as comprising incommensurable paradigms or competing ideologies. However, those who view STS as a metadiscipline concerned with metaknowledge should have no conceptual problems with this contextual shift. The knowledge and practices of scientists and engineers is, after all, what we study. There is reflexive validity in putting academic STS at the service of knowledge professionals.

Why then should knowledge professionals take to STS? We have already seen the answer to this question: There is a distinction to be made between the disciplinary sciences and communities of knowledge workers. The political and commercial worlds are in a period of transition from the Cold War era. Their old ways of doing things are proving less and less successful in a new, global political and economic context. As represented by writers such as Reich and Drucker, they are looking for new paradigms for the practice of innovation, new models of how to go on, of how to manage. Even the academic science communities, which depend on economic and political support from government and industry, are affected by these broader societal changes. At a time when academic researchers, and even federally-sponsored technology developers, are called to account for their ability to respond to the needs and values of other, broader communities, there is little to fear from ideological attacks from the Scylla of positivistic science, as long as STS avoids the Charybdis of strong relativism. The workplace, of course, has an antidote for this latter pathology, namely the marketplace, an arbiter for which all practices and all problem solutions are *not* of equal value.

There will be STS scholars who believe providing academic services to knowledge professionals is both conceptually and socially feasible, but who object to the marketplace as arbiter on ethical or political grounds. The market for professional services is a broad and complex one. It includes public interest groups, political parties, religious institutions, charities, and advocacy organizations as well as business firms, the Department of Defense, state government agencies, and universities. Knowledge professionals work in these sites and many more. The market for the professional services of knowledge workers is broad but not monolithic in its values or ethics.

The model of the scholar speaking truth to power is one which STS research itself calls into question. We believe truth to be elusive and in some ways dependent on social context. We know that power must be obtained in order to influence power. The informed participation in the workplace by critical professionals working for different interests provides an excellent and nonexclusive model for influencing the shape of society which may be more effective than either lectures from the ivory tower or grass-roots democracy. In the modern world, power is distributed across federal, state and local government, large and small business firms, elite and land grant academic institutions and a bewildering variety of nonprofit institutions, all of which play many roles in planning for science-based innovation. Knowledge professionals comprise a significant, widely-distributed, and growing part of the public who is engaged in decisionmaking regarding technological futures on a continuous basis. Their education should be of special interest to the STS community.

STS GRADUATE EDUCATION FOR KNOWLEDGE PROFESSIONALS

What then, does an advanced degree in STS offer the knowledge professional who aims to be a "manager who makes knowledges productive"? Our initial direction may be taken from Drucker, when he said that the essence of management is not techniques and procedures, but a social function which in its practice is truly a liberal art. If we are not primarily concerned with the teaching of techniques and procedures, then what is our concern?

Most of the literature on STS education since the 1960s has been concerned with the education of STS scholars, of scientists, or of the general public (Edge, 1995, 8–15). The emergence of the knowledge professional brings a new dimension to the discussion, but many of the issues are familiar. In his overview of the field, "Reinventing the Wheel," David Edge notes that one wellspring of STS in the 1970s was a concern within European universities to provide scientists with a broader education to compensate for a curriculum that was too specialized and illiberal (p. 8). The goal was to acquaint science, technology and engineering students with aspects of society in which they would work. It is noteworthy, however, that Edge quotes Clive Morphet's characterization of these reforms as "non-vocational elements" of the scientist's education (p. 9). For knowledge workers, although they may have been initially trained as scientists, a knowledge of the society in which they work has become a definite vocational concern.

Edge identifies a second area of historical interest to STS educators: the training of policy analysts and policymakers. Derek de Solla Price's *Little Science, Big Science* (1986) is the icon for this strand of STS which deals with themes familiar to those concerned with the efficiency of investments in research and development in the post-Cold War era. Although hardly a mainstream element of STS education today, this approach, which Edge labels *technocratic* (6–7, 12) can still be found in practice, for example in the work of Francis Narin which seeks to discover the economic impacts of investments in academic research through a bibliographic analysis of patent citations. Edge places techniques such as cost–benefit analyses in the same technocratic category. The application of such quantitative social science methods to the analysis of science policy issues is a legitimate activity, but a specialized one. This kind of practice is one of the specialties that a knowledge professional might manage, broker, and apply to problem solving, but, like urinalysis to Drucker's physician, they are at most a tool in the knowledge professional's black bag.[2]

Knowledge professionals, as previously described, would not be well served by a technocratic education. As Drucker pointed out, techniques and procedures are not the essence of the profession. Rather, the knowledge professional must orchestrate the performance of knowledge specialists to make their knowledges productive. The knowledge professional performs a social function which Reich has further characterized as the brokerage of the solution of problems—problems concerned with science-based

---

[2] Not to be confused with the infamous black box.

innovation. Orchestraters and brokers are not well served by technocratic methods. Rather, they are self conscious participants in the creation of new knowledge whose integrative role requires a reflexive awareness of their practice. They must adopt a critical perspective on the knowledge and practices of their specialist collaborators. Like a systems engineer, they must seek to design a solution that "works" where the judgement of what works is shared by an external community as well as an internal one.

What Edge describes as the "critical approach to science education" is thus of greater relevance to the knowledge professional than is the technocratic.[3]

> ... the "critical" approach implies teaching material about science, its institutionalization and social structure, its values and practices, so as to stress *ideas* about its social nature and its relationship with other social institutions; these ideas are taught in a reflexive manner. ... In other words, the educational goals of teaching in this mode are essentially *cognitive*; the aim is understanding of the social roles that the graduate might eventually assume, and of the institutional pressures implied by those roles, linked with the self-awareness that will activate these roles creatively. The status and credibility of technical expertise, and the justification of those social roles are *not* taken for granted. (14)

My experience as a teacher supports this argument. The M.S. curriculum offered to working knowledge professionals at the Northern Virginia Center is the same as that offered to future Ph.D. students in Blacksburg. It has a strong core emphasis on history, philosophy, and the social sciences, which comprise 60 percent of the program. With 20 percent devoted to a thesis, this leaves only 20 percent of the curriculum for elective courses which can be interdisciplinary STS courses, or courses in a science, technology, management, or policy discipline. This curriculum has attracted and satisfied a solid group of students who believe it to be relevant to their needs.

And why should this not be the case? The curriculum addresses the nature of knowledge systems: It surveys epistemology and the philosophy of science from the Vienna Circle to the present, with some attention to the philosophy of experiment and of technology. There is an emphasis on post-Kuhnian, historical-based philosophy of science and social epistemology. It considers the sociology of scientific knowledge including the Strong Programme, and questions scientific epistemological exceptionalism. It includes the considerations of Staudenmaier and historians of technology regarding the relationship between science and technology and the independent status of technologies as knowledge systems. In short, the curriculum provides a sound, liberal education regarding contemporary understanding of

---

[3] Edge is writing of the education of scientists, but I think he has in mind the roles played by scientists working as knowledge professionals.

the nature of human knowledge. It shows how knowledge is resident in and validated by knowledge communities, and provides an intellectual basis for performing the social role of making knowledge productive.

The curriculum also gives examples of how the production of scientific and technological knowledge has been influenced by social factors, according to a variety of STS scholars and historians of science and technology. Students read canonical STS works by Shapin and Shaffer, Latour, Constant, Cowan, Hughes, Pinch, Law, and Bijker among others. They receive an introduction to archival research and are required to write a historical study in science or technology, using archival sources, an exercise that students seem to enjoy even though few, if any, will become historians. In addition to providing an historical context for their work, this part of the curriculum provides potent lessons on the values and limits of historical case studies as well as firsthand experience of how historical studies selectively depend on surviving documentation. It also provides a palette of critical methods for later use.

The study of historical examples of problems in innovation provides vicarious experience to knowledge professionals. Reading critical analyses by STS researchers can also suggest frames for critical analysis and help identify potential problem areas. The curriculum is meant to engender a healthy skepticism regarding positivistic or deterministic interpretations or approaches that offer simplistic or monolithic solutions to complex problems. Students read the works and analyses of social scientists from across the spectrum, from Merton to Bloor and from Winner to Rosenberg. They are exposed to a variety of critical perspectives, and encouraged to consider their own.

The graduate seminar is our classroom model, where students are expected to read critically and make classroom presentations about their reading, while preparing an analytic paper as a final product for the course. Students may choose a work-related topic for their analytic papers. On a few occasions, paper have served dual duty, and have been submitted both for a grade and as a deliverable at work. More often, students use a seminar to explore a research site or an analytical framework generally related to their work or career interests. Thus, if they choose, students may write six different papers which address related areas of technical or social interest to fulfill the requirements of six core courses which emphasize the philosophy of science and technology, the history of science and technology, and social science approaches to the study of science, technology, and society. They may then write a thesis which draws upon their prior work. Examples include a student whose papers dealt with some issue or problem related to the emerging field of nanotechnology and another student whose papers dealt with issues of manned space exploration. Their theses, in preparation, deal with an analytic comparison of centralized versus market-driven models for the development of nanotechnologies, and a study of official and public approaches to evaluating risk and in making decisions in the light of such evaluations, in the launch of nuclear powered space probes. It is worth mentioning that, in addition to STS majors, graduate students in public administration and policy, education, computer science, and engineering have found it worthwhile to take one or more of our core courses as electives.

The faculty for this program which seeks to meet the needs of knowledge professionals includes five working, knowledge professionals. Two have doctoral degrees in STS (both history-oriented). Three are broad-base social scientists. Their doctorates are in anthropology, economics, and sociology. One is a distinguished STS scholar, the others have been attracted to the field relatively recently. These faculty members work for the National Science Foundation, the Institute for Defense Analyses, the Renewable Energy Project, Battelle Pacific Laboratories, and the American Association for the Advancement of Science. The sixth is a philosopher who splits his time between Virginia Tech and George Washington University. In addition, we have experimented with teleconference technology and have conducted courses linking classrooms in Northern Virginia and Blacksburg. Two philosophy courses have originated from Blacksburg, and an advanced course in science policy originated in Northern Virginia.

The Virginia Tech STS program in Northern Virginia is a new one; it is now in its sixth semester. Four students have already expressed a strong interest in continuing toward a Ph.D. Three of these students are not interested in preparing for an academic career, but in continuing study to support their career as a knowledge professional. There is no question that, at both the M.S. and Ph.D. level, earning the academic credential is seen to be an important career step. In the three years of the program, several students have changed jobs or secured promotions where, they report, their progress toward an M.S. in STS has been an important factor in their successful career move. Senior positions in the knowledge professions may require a Ph.D. or are, at least, more accessible to Ph.D. holders. Two of our faculty members have secured such positions based on their doctorates in STS, and our students see this as a model for their own careers. As Drucker (1994) predicted, formal education is important to the knowledge society, and students are seeking to acquire advanced knowledge "... well past the age of formal schooling" and "through educational processes that do not center on the traditional school" (p. 9). The only untraditional aspect of the Virginia Tech program is that classes are all offered in the evening, meeting once a week for three hours, so that working professionals are able to attend. It is a tribute both to the values and the energy of the students that

they faithfully pursue such a program. It is also a testimony to the value
of graduate education in science, technology and society studies to the
critically-inclined knowledge professional. It is worth mentioning that, as
part-time students, these working professionals are not eligible for fellowships or assistantships. They either pay their tuition themselves, or persuade
their employers to do so.

An STS degree does not appeal to all knowledge professionals, of
course. More than five times as many students are enrolled in the M.B.A.
program and in the M.S. programs in computer science and engineering.
Three times as many are in the M.P.A. program. But these well-established
programs have the advantage of longevity and familiarity to potential students. Much of my time with inquiring students is spent explaining what
STS is about. They all know or learn that, unlike an M.B.A., or an M.S.I.E.,
an STS degree does not carry with it a socialized career path. Those who
pursue the program are persuaded of its value which depends, in part, on
their ability to broker the degree in the marketplace. Nonetheless, with
apologies to Imre Lakatos, it is a progressive program. Each year sees more
students than the last. We have enrolled our first students who learned of
the program from other students. We have enrolled our first students who
were specifically looking for an STS program, and who found us by way of
our web site. It seems clear from our limited experience that the principal
barrier to increasing enrollments is a lack of contextual knowledge of STS
in the workplace. This is a barrier that is surmountable in individual cases
in the short term, and which is likely to be reduced over time as our students spread throughout the workplace. Evangelism by the STS community
as it reaches out to potential students would also reduce that barrier. So
far, we have done very little in the way of outreach to promote the idea of
STS education for knowledge professionals, and our advertising budget is
limited to three small ads per year in the *Washington Post*.

FROM GLIMPSING ONE TREE TO SURVEYING A FOREST

I have tried, in this chapter, to build a general case for STS education of
knowledge professionals, based on a consideration of societal trends in the
workplace as described by Robert Reich and Peter Drucker, as interpreted
through my own experience—first, as a knowledge professional, and then
as director of an STS program aimed at working professionals. I have drawn
briefly upon Steve Fuller and David Edge's characterizations of the STS
community and the history of educational themes in STS. I have also drawn,
vicariously, upon the experience of other faculty members in the Northern
Virginia program. Missing here is a consideration of the experience of other

universities with the graduate education of knowledge professionals. It does not appear to be a topic that has been widely considered. In the absence of a developed literature, I conducted a brief survey of web-based information, followed up, in some cases, by telephone conversations with faculty or administrators. The following section is not at all definitive, but strictly exploratory.

As an index to programs, I used the web sites at North Carolina State University, entitled "Science, Technology, & Society Programs at Other Universities," (Hamlett, 1996), and the AAAS web site, "Guide to Graduate Education in Science, Engineering and Public Policy" (1997). There are a total of 32 United States colleges and universities offering programs in STS or in science and technology policy listed at these web sites. Of ten found only on the North Carolina State site, five appear to be strictly undergraduate programs. None of the other five appear to offer terminal master's degrees of the kind discussed in this essay. Of the ten found only on the AAAS website, all but one, Eastern Michigan University, appear to be predominantly public policy programs with a science or technology option. Several of these 19 universities' programs may warrant further investigation and characterization, but from a cursory examination, they do not appear to have a primary concern with a graduate-level, multidisciplinary, critical approach to science and technology in society.

Of the 12 universities found on both sites, two, Carnegie-Mellon University and the University of Pennsylvania, do not offer a terminal master's degree. Georgia Tech and MIT each offer two programs, one of which is policy oriented and one of which is liberal arts oriented, but the MIT program in STS is a Ph.D. program that does not seek master's degree candidates.

This winnowing of the web sites thus yields twelve interdisciplinary programs which might offer master's degrees to science-based knowledge professionals. Five of these are most readily categorized in the liberal arts tradition:

1. Cornell University's M.A. in Science and Technology Studies
2. Eastern Michigan University's Master of Liberal Studies (M.L.S.) in Technology
3. Georgia Institute of Technology's M.S. in History of Technology
4. Rensselaer Polytechnic Institute's M.S. in Science and Technology Studies
5. Virginia Tech's M.S. in Science and Technology Studies

Seven degree programs are more closely identified with public policy approaches:

1. Georgia Institute of Technology's M.S.P.P. program
2. The George Washington University's M.A. in Science, Technology and Public Policy
3. The University of Minnesota, Hubert H. Humphrey Institute's M.S. in Science and Technology Policy
4. Princeton University's M.P.A. in Science, Technology, and Environmental Policy
5. The University of California at Berkeley's joint M.P.P.—M.S. program in the Goldman School of Public Policy
6. Syracuse University's M.P.A. (focus on technology) in the Maxwell School of Public Affairs
7. The Massachusetts Institute of Technology's M.S. in Technology and Policy[4]

There may, of course, be other programs not captured by these two web sites.

Available material on the policy-oriented programs supports the idea that science and technology policy practices are closely related to those of a broader group of knowledge professionals. The George Washington University entry at the AAAS web site, for example, states, "Graduates of the science, technology, and public policy program have had continuing success in locating exciting policy-making, research and management positions in both the public and private sectors" (no page number). The Georgia Tech entry says, "About half of the M.S.P.P. graduates find work in government agencies and half work in private or nonprofit sectors. Most find jobs as policy analysts in government or consulting firms. Recent placements include: Jet Propulsion Laboratory, Council on Competitiveness, Southern Technology Council, Georgia Power Company, the U.S. Environmental Protection Agency, the Georgia Conservancy, Motorola, Intel, and the U.S. Department of Energy" (no page number). M.I.T. reports: "The Technology and Policy Program equips its graduates with skills that may be applied to careers in the public or private sectors. The graduates work about half and half for the private sector and for government organizations" (no page number). Other policy-oriented programs make a similar claim, except for the program at U.C. Berkeley, which has only recently accepted students (AAAS). The fact that a large number of graduates from science and technology policy programs are finding work in business firms and nonprofit organizations provides evidence that the workplace is creating positions for

---

[4] Preliminary information on all of these programs may be found at the cited AAAS web site, which has links to program web pages for all programs. The AAAS site is maintained and updated annually.

vocationally mobile knowledge professionals. Of the policy-based programs, it is clear from their web-based materials that Georgia Tech, the University of Minnesota, and George Washington University have a significant number of part-time students who are working professionals. The other programs do not mention part-time students.

The major difference between the policy-oriented programs and the STS programs appears to be in the critical disciplines which they combine. Policy programs have more of an applied-bent, that is to say a greater presumption that graduates may work in other than academic positions, but this presumption is not of major significance for students seeking a terminal or vocational master's degree, since that degree is not sufficient to pursue an academic career. Policy programs typically include courses with a disciplinary home in history, political science and, economics. STS programs typically include courses which have a disciplinary home in history, philosophy, sociology, and political science. There are surely differences in practice as well, which must be reflected in curriculum content. For example, policy programs probably emphasize global and national issues where STS programs probably pay more attention to smaller communities of scientists and engineers. But the underlying critical approach and analytical skills are similar. I have found that students in the Public Administration and Policy Analysis program who take STS courses at Virginia Tech do as well in class as do the STS students.

Available material on the STS-oriented programs also supports the idea that there is a market-driven demand for higher education of science-based knowledge professionals. The Cornell program has had little experience with or demand for terminal master's degrees (Dear, 1998). At RPI, on the other hand, as on the main campus of Virginia Tech, about half of the students are pursuing a "terminal" master's degree in STS, although some of these may go on to advanced degrees in other fields. RPI offers an internship option to the thesis for M.S. students. Recent graduates have found jobs in consulting, in public interest advocacy, and in science and engineering management (Vumbacco, 1998). The Georgia Tech program in History, Technology and Society also has a strong orientation to knowledge professionals. Consider the following statements taken from the program's home page (1996):

> An HTS degree can provide a solid profession for almost any career. Historical and sociological methods, skills, and data are used in virtually every profession, including business, social services, journalism, and government. . . . Employment analysts predict that today's college students will change careers—not just jobs— three to four times in their working lives. And individuals with narrow pre-professional degrees are likely to find that they do not have the intellectual agility to achieve their professional goals. They find themselves passed over for

promotion by colleagues who have broad knowledge, analytical skills, and the ability to communicate their ideas forcefully in written and spoken form. . . . At the recent Ivan Allen dinner at the Ritz-Carlton, an executive working with Tom DuPree reflected on these facts in a discussion of his own career. He noted that it was not his education as a mechanical engineer that prepared him for his work with DuPree, but the general education courses in arts and sciences that he took along the way. Indeed, he believed that the greatest challenge at Tech is to encourage students to obtain a broad education, rather than one that is narrowly vocational. That is what an HTS major is all about. (no page number)

Finally, two of the STS programs are specifically aimed at educating working knowledge professionals. They offer evening classes, and their students are primarily working professionals. They offer both thesis and nonthesis options. These are Virginia Tech's three-year-old, STS graduate program at the Northern Virginia Center and Eastern Michigan University's well-established Master of Liberal Studies in Technology, in the Department of Interdisciplinary Technology. Although Eastern Michigan's program departs somewhat from the STS curriculum offered at the other schools in its emphasis on technology and its large number of electives or complementary concentrations (Eastern Michigan, undated), the program includes both philosophy and sociology of science and technology, and is designed to produce a critical perspective (Hanewicz, 1998a). The program serves nearly 400 students, the large majority of whom are part-time, working professionals in a large variety of fields. Wayne Hanewicz, the Graduate Program Coordinator believes that the program, which has tripled in size over the last six years, is serving an educational need which is generally unrecognized and underserved nationally.

SOME CONCLUSIONS

The importance of science-based technological change to human societies and civilization has been an accelerating characteristic of the 20th century. It is a complex phenomenon that has attracted critical scholarship in a number of disciplines and has provided an object of study for interdisciplinary scholarship, leading to the establishment, in the last 30 years, of academic programs in science, technology, and society studies and science policy. At the same time, the workplace has evolved in such a way as to create a demand for science-based knowledge professionals. These people manage the course of technological change and provide critical analyses in support of planning for such change in all social sectors, including government, industry, academia, and a variety of nonprofit institutions. Knowledge professionals comprise an important and growing part of the workforce in the post-industrial age. They are highly educated, technically-oriented people, whose midcareer jobs require them to identify, solve, and broker the solutions to problems of science-based innovation. Individually, they make important decisions, and, collectively, they create the trajectory of our technological and social future.

The connections between those who practice the knowledge professions and those in the academy who study science and society are, with some exceptions, weak. A few knowledge professionals have found their way into STS graduate programs, but for the most part, graduate programs in STS have not recognized this potential constituency, and most knowledge professionals have never heard of STS. Science policy programs, with a weaker bias toward academic careers, have attracted knowledge professionals with an interest in government, but the federal government sector is only part of the workplace for knowledge professionals, and one of declining relative importance. As a consequence, perhaps, as many as half of the graduates from science policy programs are finding work outside the government sector, a fact which has important implications for their curricular content and for the way such programs are conceived and marketed to potential students.

The arguments of Drucker and Reich indicate that knowledge professionals are in need of critical analytical skills and professional knowledge which they can use to broker solutions to problems which involve both technical and non-technical communities. As their job responsibilities increase, from knowledge creation in an initial disciplinary or application's task, to the management of more complex projects, they need to learn how to assess and integrate the knowledge produced by multiple communities of specialists and new technology users. Although most practitioners would reject the term as esoteric, knowledge professionals must function as practicing social epistemologists.

Yet, the STS education of knowledge professionals has typically been on-the-job training. They learn how to manage complex tasks by observing others. They learn from experience that positivistic approaches to problem solving and the generation of new technological knowledge do not work. They engage in the social construction of new technologies as a form of practice, tacitly acquired. Their higher education has typically focused on technical knowledge or on the tools and procedures for decision making—advanced degrees in science, engineering, or management, although, according to Drucker, these are not the critical skills which knowledge professionals need.

Science, technology, and society studies is the academic field concerned with the critical analysis of the creation, development, dissemination, and validation of new scientific and technological knowledge. Other than self replication, however, STS educators have been focused on the general, nonvocational education of scientists and engineers, on the one hand, and on general education to empower the public in dealing with science-based technological change, on the other. These are worthy goals, but STS educators have largely missed the point that the public does not deal with communities of scientists and engineers directly, but through knowledge professionals—knowledge brokers, working in positions throughout the workplace—in government, in public interest organizations, in business, in the media, and elsewhere. If Drucker is right, and if the growing experience of urban schools like Eastern Michigan, George Washington University, Georgia Tech, RPI, and Virginia Tech's Northern Virginia Center are any indication, then there is a latent need and a current opportunity for the STS community to provide graduate education aimed at preparing knowledge workers for nonacademic careers.[5]

Moreover, the emergence of the knowledge professionals, as documented and discussed by authors such as Drucker and Reich, is an aspect of science and technology in society which has largely been overlooked by researchers in the STS tradition. Their professional practices as knowledge brokers concerned with managing technological innovation through the integration of scientific and technical knowledge on the one hand and social knowledge (managerial, institutional, cultural, and political) on the other, can provide important research sites for scholars interested in understanding the ways in which new knowledge is created, negotiated, and disseminated in our complex society.

We in the academic STS community should apply our considerable analytic methods and critical insights to better understand the practices and the social roles of knowledge professionals. We should examine the ways in which our traditional curricula in epistemology and the social sciences relate to their practices and professional needs. We should strongly consider reforming our educational programs to attract their attention and to meet their needs.

REFERENCES

American Association for the Advancement of Science (1997). Guide to Graduate Education in Science, Engineering and Public Policy. http://www.aaas.org/spp/dspp/sepp/index.html.

---

[5] I don't mean to exclude careers in academia. University positions in the office of sponsored programs, in technology transfer and in research administration are among the jobs filled by "nonacademic" knowledge workers. If the academy would lower the career barrier to nontenure-track professors, the distinction could disappear.

Dear, P. (1998). Telephone conversation. February 6, 1998.

Drucker, P. F. (1994). The age of social transformation. As originally published in *The Atlantic Monthly* (November), http://www.theatlantic.com/atlantic/election/connection/ecbig/ soctrans.html. [1998, January 6].

de Solla Price, D. J. (1986). *Little Science, Big Science . . . and Beyond.* Columbia University Press, New York.

Eastern Michigan University. Undated. Master of Liberal Studies in Technology. http://emich.edu/public/emu_programs/grad/cot/idt/mls.html [1998, January 30].

Edge, D. (1995). Reinventing the wheel. In Sheila Jasanoff, *et al. Handbook of Science and Technology Studies*, Sage Publications, Thousand Oaks, California. pp. 3–23.

Fuller, S. (1997). Constructing the high church—low church distinctions in STS textbooks. *Technoscience 10:3* (Fall). http://www.cis.vt.edu/technoscience/97/fall/ 97fall. htm, [1998, January 18].

Georgia Tech. School of History, Technology and Society (1996). What Can I Do with and HTS Degree? http://www.gatech.edu/hts/general/what.htm [1998, January 30].

Hamlett, P. (1996). Science, Technology, & Society Programs at Other Universities. http://www2.ncsu.edu/ncsu/chass/mds/stsprog.html [1998, January 30].

Hanewicz, W. B. (1998a). Graduate Program Coordinator, Eastern Michigan University, Department of Interdisciplinary Technology. Telephone interview. February 4, 1998.

Hanewicz, W. (1998b). Telephone conversation. February 5, 1998.

Narin, F., Hamilton, K., and Olivastro, D. *et al.* (1995). Linkage between agency-supported research and patented industrial technology. *Research Evaluation* 5,3: 83–187.

Reich, R. B. (1991). *The Work of Nations*, Vintage Books, New York.

Vumbaco, K. (1998). Telephone conversation. February 4, 1998.

Webster's Ninth New Collegiate Dictionary (1985). Merriam-Webster Inc., Springfield, MA.

CHAPTER 10

Reculturing Science
Politics, Policy, and Promises to Keep

Daryl E. Chubin

This is a blend of reminiscence, policy analysis, and political commentary. I shall draw on my experience in and out of the federal government and try to temper both self-righteousness and cynicism as I describe, and then prescribe, what "science watchers" in Washington D.C. and its surrounding reality, fear, loathe, and cope with daily. The epigram for my remarks might be Daniel Bell's observation that an academic is one who wonders whether something that works in practice will work in theory. The tension inherent in Bell's remark captures the tension that Nicholas Mullins will always symbolize in my mind: He was my competitor, colleague, and friend.

Our relationship was intertwined with 4S, the Society for Social Studies of Science. Indeed, the chapter we coauthored for a 1985 collection on *Managing High Technology* (Mullins and Chubin, 1985) was titled "Society for

Daryl E. Chubin, National Science Board Office, National Science Foundation, Arlington, VA 22230.

A version of this chapter appeared in *Science and Public Policy*, February 1996. Copyright 1996 by Beach Tree Publishing. Reprinted with permission.

Science, Technology, and Society: A Sourcebook on Research and Practice, edited by Kumar and Chubin, Kluwer Academic / Plenum Publishers, New York, 2000.

social studies of science: perspectives on interdisciplinary research." In it, we cite collaborations, respectively, between Latour and Woolar, Edge and Mulkay, and Griffith and Small. Then we add "work on interdisciplinary subjects such as the discovery and growth of reverse transcriptase (which each of us has done independently and with divergent results . . .)."

It was the future of 4S that brought us together. The society was formed in response to the discipline bound constructions of the role of science and technology in society. As co-founders Nick and I were part of an alliance of scholars on both sides of the Atlantic who drafted a charter asserting the need for new paradigms, syntheses, and expectations of professional practice. Nearly 20 years later, and almost a decade after Nick's challenging, even sullen, presidential address in Ghent, the 4S charter remains largely unfulfilled.

One reason for this, I believe, is Nick's absence. Today, 4S is an amalgam of mostly academic scholars who represent a research specialization that rivals disciplines but falls short of an "interdiscipline"; who are more exclusive than inclusive in orientation (they discourse about gender, ethnicity, nationality, and "difference" but are still self-referential and mostly hierarchical in their claims); and who have yet to translate the value of 4S into the education and mentoring of new cohorts who will enter various sectors of national work forces to effect change.

If Nick were with us still, I think he would caution that "research is not enough; careers in academe are but one choice; interacting with other 4S members is the easiest thing we do." Nick might even be practicing in Washington, nodding in approval if I asserted the need to connect with other constituencies—journalists, elementary and secondary educators, those knowledgeable about industry, local and national policymakers—to refocus 4S thinking and interests. As a society, 4S is as obligated to communicate with those who *don't* know and *don't* care about the problems and solutions inherent in science and technology as to confer in structured public fora with stakeholders already invested in the processes and outcomes which they manage.

After 14 years as a university faculty member in which I, like most 4S members dissected the culture of science, I moved to the Office of Technology Assessment (OTA) to do policy analysis and advise Congress. That career lasted seven years, and now I am seven months into serving the executive branch at the National Science Foundation (NSF), where I oversee a portfolio of activities that, simply put, intervene in the status quo. I am also responsible for evaluating the difference those activities are making "in the field." I have the opportunity to act on the very advice I gave NSF years before and to be accountable to various publics for the performance of the portfolio.

Yet my experience and career path is not unusual. That, too, keeps me mindful of Nick, for it was our mutual concern for kids, soccer, and "careers" as a subject for social analysis that kept us together. Through our time at, and affection for, Cornell University, our independent adoption of Derek de Solla Price as an informal mentor, and our intrigue with policy, we became what Nick called "trusted assessors." Today, I continue to teach public policy part-time to impart an STS (science, technology, and society) world view, to draw students to public service, and to develop, I hope, a new generation of trusted assessors.

MAKING CHANGES

The title "Reculturing Science" derives from the oft heard phrase—at least where I live—"reinventing government." Like many exhortations, it passes the lips all too effortlessly. Making changes happen is another matter. But happen they must. As a nation, a government, a citizenry, we cannot afford to do otherwise. Besides, we are too smart *not* to reinvent, reorganize, and reinvigorate ourselves. Our institutions demand it and our ingenuity as a workforce is up to the task.

Habits, however, are hard to break. The force of tradition is potent because it is familiar, reassuring, and predictable. Inertia, distrust, and risk-aversion are our foes. As the Clinton administration is apt to point out, there is a crisis of confidence in government. The level of cynicism, according to the polls, is higher than after Watergate. The public, half of whom do not vote, want change in their government to improve their lives.

In a series of surveys completed by the Americans Talk Issues Foundation, pollsters tested 50 reform proposals by telephone for a 10-day period in January 1994. The greatest consensus formed around term limits for elected officials. However, the results in general suggested that the public wants more direct access to the political decisionmaking process, including the opportunity to define the government's spending priorities.

It also wants to evaluate government agencies by the results they produce rather than the programs they initiate or the money they spend. To quote the *Washington Post* article (20 April 1994, p. A19):

> Objective scorecards would measure such indicators as how much students have learned, whether the water and air are cleaner and how many people on welfare have become self-sufficient.

As one of the pollsters concluded, there is a "real sense of the wisdom of the people."

To cap the thought: Our economy, and that with which we globally commune, requires that we reduce government. Either we do it smart or we do it dumb. Dumb has been tried: across-the-board cuts, paper work, regulations, excessive hierarchy, lack of responsibility, diffuse accountability. Reinvention is a better idea. Accountability for performance and outcomes, doing more with less, harnessing technology, "delayering" or flattening the bureaucracy, teamwork, identifying and serving the customer. This is not a paid political announcement; it is a change of ethos. It is a reculturing of government. It should be seen as a model, not an imposition.

Reculturing must occur from the inside of social institutions, otherwise it is hardly transforming, enduring, or effective. We all know change takes time, hurts a lot, and redefines boundaries, behaviors, incentives, rewards. It is what science, too, must do.

I hear you saying to yourselves, "Isn't this politics?" Yes. "Isn't it policy-driven?" Yes. "And just who has promises to keep?" We all do. The sooner that science (scientists and engineers, their professional associations, their lobby organizations, their rhetoric, and their sponsors) recognizes that change is good and that it is better to embrace than repel, the sooner universities will be happier places, students will be better served, and the federal government will be seen as a partner and positive force in providing needed interventions in society. Enlightened self-interest is an incomparable motivator, and there are ways of aligning it with what is murkily referred to as "national interest."

Let us move this interest closer and more concretely to our world of research and higher learning. Consider this quote from *The Lost Notebooks* of Loren Eiseley (Heuer, 1987): "A university is a place where people pay high prices for goods which they then proceed to leave on the counter when they go out the door." We hope not, yet we snicker knowingly because there is something to it.

A recent inquiry received at NSF from *Washington Post* science journalist Boyce Rensberger puts Eiseley's wry observation in other terms. Rensberger reports he is preparing a series to run in the paper and would like "to hear from scientists, historians of science or other science watchers who have data and/or insights on the health of American science." He poses these questions, many of which have been on my agenda, in a sustained policy sense, for the last five years:

- Is the funding squeeze really as bad as some say?
- Is it forcing good people out of science or just the marginal ones?
- Does stiffening competition for grants cause people to abandon high-risk, high-payoff ideas in favor of safe-but-dull proposals?
- Is the competition causing scientists to cut corners, or steal ideas, or do other things that are supposedly unprofessional?

- Are the pressures for more applied science and for quick payoffs hurting basic science?
- Do we have a national science policy and, if not, do we need one?
- Who should devise such a policy?

Answering such questions is a promise I have tried to keep. Let me share my attempt to address them.

Then and Now

The research community of the 1950s and 1960s enjoyed a funding situation very different from that of today. Then the research environment was characterized by:

- Fewer researchers, ample job opportunities, and a more homogeneous workforce of white males: This composition was reflected in the student population
- Fewer research universities with a greater concentration of federal resources
- Little international competition or concern about United States research performance—the United States was dominant
- A compact that entrusted the judgment of scientific merit to expert peers in return for scientific progress that would benefit the nation (national security, productivity, and quality of life)
- Researchers' expectations of sustained federal funding, which is now pejoratively called an "entitlement" mentality among scientists

That was then and this is now. I would suggest that, when funding falls short of expectations, when the number of deserving researchers is such as to deprive some of the chance to pursue promising opportunities (or pursue them as fast or as fully as they hoped), the result is a relative, not an absolute, deprivation. "Relative deprivation" is a social science concept that highlights the disjuncture between federal funding trends on the one hand, and institutional and personal angst on the other. This is what numerous speeches, surveys, and reports have captured in the last three years.

Does knowledge or this paradoxical disjuncture lessen the pain or lift the morale? Not necessarily. But there is an implied corollary question: Does the research community expect that the federal government will fund every deserving researcher? Some adjustment in thinking in the research community, in the government–university compact, is needed.

Put another way, we are in the midst of a "generation gap," a gap not only of experiences, but also of perceptions and proposed solutions.

Scientists and science educators today, all well-intentioned, may be deceiving themselves, but moreover, they may be misleading their students and protegés. They are mistreating the future.

I wish to discuss careers first from the "bottom up" rather than from the common policy perspective of the "top down." We need, of course, to keep both in sight—the aggregate trends and the individual experiences—when considering markets, recruitment, and images as factors in career opportunities for scientists and would-be scientists.

FALLACY OF THE MARKET

Employment markets are never as rational as labor economists suggest. Societies thus face mismatches in the labor force between people and jobs, between opportunities and supply, between current skills and tomorrow's demands, between specialists with great depth of understanding but little breadth and the organizational need to hire specialists but to grow generalists (e.g., Breneman and Youn, 1988).

Like all professions, especially medicine and law, that require post-baccalaureate degrees, science pays in the long run (Chubin, 1990). Yet salary is an overrated variable in predicting future supply and demand for scientists. Salary taps only one consideration in the motivational inventory of the student-cum-scientist.

Similarly, the award of Nobel Prizes to Americans is in no way indicative of the strength of the United States research enterprise. It may have to do with the sheer number of scientists working here. This, too, is a fallacy of the market, the mistaken connection of motivations with outcomes, of high positive correlation but to no known casual relationship.

In a dynamic economy, shortages and surpluses are expected, with variations by sector, industry, and region of the country. Relative size of one's birth cohort also contributes to competitive advantage and disadvantage. "Baby Boom" and "Baby Bust" describe significant demographic realities that influence expectations as well as flows into and out of the work force (Sargent and Pfleeger, 1990). The question is, how insulated from, or vulnerable to, such market fluctuations are scientists?

CALLING VERSUS RECRUITING

Science can be a multiplicity of careers. Over the life course, movement within the labor force (between sectors, organizations, and jobs) should be expected. To speak, therefore, of science as a singular vocation is as much a distortion as the notion of the "typical American." What does a scientist "look" like? What does he or she do? In what field did he or she earn a degree? These should be baseline questions, not limiting conditions.

Careers envelop people: They cause people to live them 24 hours a day. Indeed, careers in science reveal the inescapable entanglements of personal and professional lives: One informs and enriches the other. To subordinate one to the other increases the tensions that individuals must negotiate with social institutions and one another.

With the family and medical leave issue being a central part of the 1992 presidential campaign, we witnessed how American public policy lags behind that of other nations and the value system of most American families. Women have lived this tension for years. Now it is being redefined as a federal responsibility as well. Thus, we are confronting not just symbolism, but something very real to careers and very relevant to their study.

Aspirations are shaped not only by information of what a job entails, but also by perceptions of "human qualities": how welcoming and supportive it may be compared to others. People choose careers, but society also signals its choices through various formal education and training mechanisms as well as an array of informal cues transmitted by the media.

Science is a laggard in its recruiting practices, and therefore in its receptivity to students other than those who have exhibited either an aptitude for, or a commitment (the "calling") to, science. The young uncommitted, the traditionally underrepresented (women, minorities, persons with disabilities), and later, those who express an interest in something other than mainstream research or teaching careers in institutions of higher education, are seen as peripheral to what is valued by science as a profession. This is a mistake repeated because there is a single prescribed model of who can do science, and therefore, who should be encouraged to do it.

Few children feel the burning in the soul to become a scientist or engineer. Yet children are what Carl Sagan has called "'natural' scientists": They have an innocent curiosity of the world around them. A subset of these children will feel the urge and the satisfaction to tinker, to manipulate things, and to learn how they work. For most I suspect that some positives reinforcement of these stirrings must come in the form of human encouragement.

This confirms for the student that indeed he or she is being called, or alternatively, this stirring is a gift. It is good to answer the call. Without this intervention from a significant other—parent, close relative, or friend who does science or engineering for a living—the calling remains detached from ambition and the measures that must be taken to respond by choosing science as a career.

The calling myth is a strong deterrent to recruiting students to science because it perpetuates the idea that there is a critical period in which one is called and if it does not happen, then it never will. Furthermore, it denies the role of information in helping students make choices, and the full play of individual differences in the expression of those choices. Were no scientists late bloomers? Did none lack confidence in their abilities? Did all have clear conceptions of what lay ahead in their careers and what would provide challenge and self-fulfillment?

The calling is a convenient code for early identification of the best and brightest. For many future scientists, this stirring is a benchmark. For others, it is irrelevant. For both categories, recruitment is essential, but when and how it occurs is variable. An educational system that recognizes this will maximize choices; it will forgive uncertainty and not penalize the "late-bloomer" (Tobias, 1992). It will make the "science pipeline" a permeable membrane. It will make school teachers, as an OTA report puts it, "talent scouts" rather than "curricular traffic cops" (U.S. Congress, 1988).

IMAGES OF CAREERS

So on the public relations front, science is lagging. Accurate information has yet to supplant stereotypes. Scientists recognize this. In a survey of members of Sigma Xi, The Scientific Research Society, a major concern was a

> lack of public understanding of what science is or what scientists do, a condition that leads simultaneously to great expectations and great trepidations—a kind of Dr. Schweitzer versus Dr. Strangelove split. (Sommer, 1987)

Most of all perhaps, the humanity of the scientist is yet to be associated with his work. And that is another, more pervasive problem.

The scientist's image is marred in several ways: it is still more "his" work than "hers"; of knowledge "products," not process; of the scientist as specialist instead of citizen. Finally, the problem is usually phrased as the public's "illiteracy" and lack of understanding, rather than the scientist's role in creating and perpetuating the gulf between the few like themselves and the rest of the population (Hazen and Trefil, 1991; Heilbron and Kevles, 1989; Langenberg, 1991; Prewitt, 1982).

The curse of the lagging image, of course, is the disservice it renders to the next generation. Who wants to inherit an anachronism? If women cannot identify with what the scientist is or does, then science suffers from a built-in deterrent to recruitment. The best models may be the unspoken ones, the ones that send visual messages instead of verbal ones. How can we gauge the importance of having had an accomplished and articulate female director at the National Institutes of Health, or an equally accomplished African-American Ph.D. physicist lead the National Science Foundation (Angier, 1991)?

The "job" of scientist, in short, is displayed in the classroom and the boardroom, in the dean's office and on the floor of the company's manufacturing plant, on the nonfiction bestseller list and in the keynote address. The traditional portrait is changing (Latour, 1987).

THE NEW SCIENTIST

The emerging image of the scientist is prominently associated with a few popularizers. These well credentialed and proven performers seek a larger audience than fellow professionals. They communicate in magazine columns and on public television, on talk shows and other fora that reach lay audiences. Most of these scientists have considerable explanatory gifts. They are entrepreneurial, and they put a human face on what is typically hidden in the jargon of technical journals and annual research meetings.

Popularizers arouse great jealousy because they demystify the guild secrets of a discipline, and they acquire an identity that makes their words a saleable commodity. That is, scientists who find a nonspecialist, or attentive, public are rewarded with the recognition of money and credibility.

Popularizers exhibit an unswerving belief in knowledge as the key to freedom and democracy in the game of citizenship. Everyone should play. To privatize knowledge or concentrate it among a select few is to limit the exercise of exploration and choice. Popularization thus becomes a compulsion that feeds both individual need and democratic ideals. It is the essence of "tithing" the giving of oneself so others may know. It is also the height of "community," a sharing of knowledge.

That is why science literacy is so often invoked as the foundation upon which democracy must thrive. Without an informed citizenry, politicians can thwart the "truths" of science. Science literacy is a testament to what all citizens should know, or know how to find out. It is a learned skepticism toward authority and believing something due to its source rather than its demonstrable logic.

Popularizers want knowledge filtered through those who "know." They do not want us to trust leaders, because political agendas can subvert what we think about nature. In a nation of specialists, popularizers arc audacious boundary-spanners. They do the work of journalists and commentators. They link microscopic detail to macro ideas. They know how to put things into context. They challenge us to challenge others, not to trust our instincts, but to experiment with an alien idea and our own senses.

Popularizers are generalists in disguise. They ooze enthusiasm and relate weighty principles to everyday experience. Above all, they are attractive in the way they look and the way they speak. They are in touch with various parts of the culture that seem to have nothing to do with science at all (Pendlebury, 1991). That is why we need this "new" scientist.

The new scientist views careers as phases, finite and alluring, entailing not "hurdles" but opportunities to exercise and develop professional skills. To give expression to skills is a kind of self-fulfillment that cannot accurately be valued by those outside: the analysts, the pundits, or the power-brokers.

INSTITUTIONS THAT MATTER

So can we get new scientists from the same old places? Can we grow the solid specialist who aspires to wider communication, a multidisciplinary practitioner? As crucial as the individual is in science, the team, the organizational culture, informal networks, professional standards and norms—the interpersonal dimension—is more crucial. Careers are embedded in, and shaped by, these social forms. But they are more than markets, recruitment, and image. Institutions matter. The institutions that matter have traditionally had education as their business: schools, families, media (Rigden and Tobias, 1991; Pool, 1990; Oakes *et al.*, 1990).

If ever a social institution required some soul-searching and stock-taking it is the United States research university. Approximately 100–200 (depending on how one counts) universities face an uncertain future. Yet the dependence of the nation on these institutions to produce Ph.D. scientists makes the uncertainty a public issue. Who is their client? What are their missions? How can their resource base be expanded without losing core values? How can faculty adjust to a new world of new markets, a diversity of incoming raw talent, and an array of career opportunities that they themselves never had to contemplate?

Recent evidence suggests that, at the graduate level, students who try to emulate the career pattern of their mentor will foreclose opportunities and feel disappointed if not betrayed. (just ask a recent vintage Ph.D. in physics.) A rededication to undergraduate education at research universities must also become a higher priority (U.S. Congress, 1991). Generally, universities must decide what they do well and what they are willing to let other institutions do for them and the society (Schmitt, 1989; Rhodes, 1991).

My policy bias, not surprisingly, is this: If institutions are underserving the aspirations of certain segments of the student population, the federal government has a legitimate role to intervene. Intervention may mean changing perceptions of science as a career, and refining expectations or strategies to turn aspirations into realities.

At least since the passage of Public Law 85-864 the National Defense Education Act of 1958, the federal government has been a key supporter of pre- and postdoctoral students. The direct role of the federal agencies, especially NSF, NASA (National Aeronautics and Space Administration), NIH (National Institutes of Health), and DOD (Department of Defense), in expanding and sustaining the pool of talent headed for careers in science, is an impressive legacy of programmatic intervention and the provision of institutional, faculty, curricular, and student support (U.S. Congress, 1985). The federal government continues to influence career choice by matching resources to policies implemented and overseen in the national interest.

There are no benign actors in education. The future is too precious to allow singlemindedness to dominate our treatment of children, or Ph.D. projections to define career opportunities. If students are treated as human resources and not as miniature images of ourselves, we will be creating career choices—from the bottom up. Let me linger awhile longer on the policy implications of what I have sketched.

MISSIONS AND DILEMMAS

The dilemmas facing American science today and for the foreseeable future can be distilled as the following big question: How can we preserve what have been the strengths of the "federal research system" while adapting to changes in the system itself? Size, diversity, competition, and productivity are at once assets and burdens. In other words: How large an research and development enterprise (*Washington Post*, 1991).

As universities know, the missions in the 1990s are creation of intellectual property, mission, and training. Education centers on changing demographics, issues of finance, classroom culture, and institutional commitment to student achievement. These mingle with a series of faculty and governance issues: What is the proper balance between research and undergraduate teaching? And, more directly, how can the rewards for excellence in teaching and advising, that is, interventions both in and out of the classroom, be structurally improved?

With respect to graduate training, universities and departments must consider the preparation of Ph.D. students for careers beyond academic research. The model whereby a productive mentor would reproduce himself 10, 20, or 30 times over may be dysfunctional for the 1990s and beyond if the career path of the new Ph.D. is intended to duplicate that of the mentor. This is not the same as saying there are too many researchers. It may mean there are too many academic researchers, or researchers at Ph.D.-granting institutions, the cadre that relies on the federal government for research support.

Chief among the factors affecting the fortunes of universities are the perceptions and actions of the policymakers, particularly Congress. These dilemmas center on the place of research and development in the discretionary budget. The roots of the dilemma ensnare the political culture of science advising, executive–legislative gamesmanship, and the congressional budget process.

First is the conflict, under-appreciated by those outside Washington D.C., between authorization and appropriations subcommittees. The former provides oversight to the research agencies and their programs. But their will is increasingly thwarted by appropriations subcommittees faced with impossible trade-offs, such as, housing versus the superconducting super collider (SSC), in the bigger budgetary picture. In the words of the Chairman of the House Committee on Science, Space, and Technology:

> . . . scientists and politicians need to raise the level of debate about funding for science programs. They must move beyond the absurdity of trading these programs against each other. Cutting science budgets to pay for low-rent public housing or the job corps is a trap that ensnares the nation in deeper problems.

A related conflict is the growth, and growing contentiousness, of academic earmarking. As universities turn to lobbyists to appeal directly to the process of distributive politics, the perception of some is the undermining of merit. The perception of others is that the building of research capacity in those regions, states, and institutions with less tradition as (and smaller concentration of) research performers is both wise and compensatory of inequities in the federal allocation system. OTA found that both these positions lack empirical support (U.S. Congress, 1991).

PROFESSIONALIZED SCRUTINY OF RESEARCH AND DEVELOPMENT

Speaking of OTA evidence (now that the 105th Congress has defunded it out of existence): the cast of characters in the policy-making process has changed. The post-World War II generation of science policy advisors transferred their experience and halos as scientists (mostly physicists) into the political arena. They were "cold warriors," many of them participants in, or direct descendants of, the Manhattan Project.

This is not to imply a single-mindedness, but instead an approach to problem-solving shaped by living the wartime-to-peacetime transition. It was an era for celebrating the fruits of democracy. An attenuation of the influence of physicists in policymaking has been openly suggested; some think that biologists will succeed them. This may be no better.

I am hopeful for more diverse participation that will change the character of the process. A new generation of professional policy analysts, some with degrees in policy and social science, is succeeding the original science policy architects and advisors. Some have migrated to policy careers. Their expertise, in short, is different from their predecessors.

To some this signals less authority for science in politics. It certainly suggests that the character of science policy will change in the current era. This has already occurred: "Professionalized scrutiny" has been institutionalized in the federal research and development apparatus at the Office of Science and Technology Policy (OSTP), NSF, NIH, and the congressional support agencies (notably the Congressional Research Service). In addition, the nonprofit stakeholders, including the National Academy complex, omnipresent think tanks and foundations, and oddities such as the Carnegie Commission on Science, Technology, and Government, and the Howard Hughes Medical Institute, constitute a reservoir of analysis and advice for policymakers.

All of this has upped the ante for information, and underscored the potential role for data in policymaking. It also shifts the discourse away from the authority of credentialed scientists, many of whom have been honored with Nobel Prizes and other trappings of impeccable judgment, and toward data-laden criticisms of what is known versus what gaps and uncertainties exist in the empirical baseline on the federal research and development system. This can expose scientists as partisans who favor their subjective experiences in the research trenches against national trend data (Chubin, 1994).

There is the rub: it is the "outsider" policy analysts who act "scientifically" in scrutinizing the scientists' support system, whereas the scientists advocate increased federal support based on their privileged "insider" status as participants in a research community. Such an appeal derives from the expectations of the longstanding government–science compact. Instead there are heightened calls for accountability (from both the White House and Congress).

Another consequence of professionalized scrutiny of research and development is that it tends to insulate policymaking from grassroots influences and vests it in a multitude of federal and nongovernmental organizations that function as a kind of "knowledge elite." Regardless of a conviction that spokespersons of research communities or policy analysts either confirm or disrupt the biases of policymakers, it is fairly obvious that ordinary citizens have a difficult time participating in this democracy of credentialed expertise. As I have already indicated the public is restive, science "popularizers" are disparaged, and science journalists reviled by researchers who nonetheless continue to lament an elusive "public understanding" of science.

If science is to reculture itself, both watchers and critics of science will have to flourish.

To recapitulate: money alone will not solve the problems of the federal research system. But the federal government, the executive and legislative branches alike, was never organized to monitor and manage the pluralistic, decentralized R&D enterprise. If they cannot link R&D funding to national goals, they will simply not look after the investments in a coordinated and flexible manner. The portfolio will continue to resemble an ad hoc collection of pet programs and projects that muster key support from key participants at propitious points in the policy-making process. The nation deserves better: a reinvented government–university compact.

For the federal partner, this issue is the maintenance of the science base, which it is struggling to do, or a greater niche-filling role that recognizes the preeminence of industry and the states to support research and development with one eye on the economy and the other on education. The agencies will need to clarify their missions instead of taking on new ones, and acknowledge that resources will force choices. Taken together, such tradeoffs would give the federal government a more specialized posture toward research and development.

This posture will either reward research universities by concentrating research and development funding in the top 10, 20, and 50 that historically have produced the most research and trained the lion's share of Ph.D.s, or distribute federal resources more widely and according to criteria that augment scientific merit. Because science and engineering are capital-, instrument-, and labor-intensive pursuits, they are elitist. Not every institution can provide the facilities and expertise to do cutting-edge research.

The federal government then could be boldly interventionist or laissez-faire in its support patterns. It could champion strategic planning and priority-setting, both for itself and for the institutions that perform research. For every research dollar awarded even more strings could be attached.

Universities, too, have several choices to make. If they find the federal government an overbearing patron, they could continue to wean themselves from federal research funding. However, most research universities' operating budgets are so dependent on federal monies that weaning is unlikely. "Leveraging" federal monies with state, private, philanthropic, and international contributions is routine, but the strings that these patrons attach may be even more restrictive, short-term, and anything but value-neutral.

So who is to lead the universities to a new relationship with the government? Few current or former university presidents and chancellors—

Frank Rhodes, Donald Kennedy, Derek Bok, Jim Duderstadt, Don Lagenberg, and Harold Shapiro may be the exceptions—have asserted the need to reinvent the university.

ENTERPRISE AS SYSTEM

Indeed, the most throughgoing challenges to conventional thinking about the government–university compact, and science policy more broadly, have emerged from a "systems" thinker who has experience in industry and government and who now directs an academic public-policy program, and a chemist-cum-university vice president for research who is now a college president. The former is Lewis Branscomb, the latter Linda Wilson. Here, in capsule form, is the crux of their challenges.

In a 1989 hearing held by the U.S. Senate (1989) Commerce, Science and Transportation Committee (and chaired by then-Senator Gore) Branscomb asserted that:

> A lot of the way human society has used science and technology really derives from a sort of tradition of linear Newtonian mechanical thinking . . . [S]ome new ideas . . . come straight out of the study of ecology . . . Ecological thinking is different from linear thinking . . . And in the past whenever we have approached the general subject of science and technology policy, we've done so within the same intellectual framework that produced the scientific revolution [that led to] many of the problems that we face today. If we could get our way of budgeting R&D in the government defined in terms of principles that express where the benefits are going to now, then we would redistribute our emphasis . . . [T]he scientific information system serves scientists very well. But they don't communicate terribly effectively with the people who can use science."

To social scientists, Branscomb's thinking underscores the historical compulsion to consult scientists about science policy and ignore specialists who examine the structure of social institutions with which science, and policies for promoting and governing it, must interact. This is where Wilson comes in. Although she is not a social scientist, her sensibilities reflect acute ecological thinking that unites the big picture with local interpersonal and institutional needs.

Wilson (1992) states that the United States research university enterprise

> was designed when most of the players were males, wage earners living near their workplaces, and supported physically, emotionally, and socially by their families . . . The design features of our academic research enterprise—from tenure to research-support mechanisms, mentoring, and information-exchange strategies—were built on assumption of a contest that permitted single-minded devotion to task, rather than a context for persons with multiple competing

responsibilities spanning work and home. What has changed is the diversity of
the roles of the players in the enterprise itself—where faculty are educators,
researchers, entrepreneurs, policy advisors, peer reviewers, public relations managers, financial managers, and personnel managers—and in the diversity of roles
played outside the enterprise.

    ... We need to tap the entire pool of talent to strengthen, replenish, and renew
our science and engineering work force. But the new "immigrants" to the science
and engineering work force, women and minorities, bring some differences in
expectations, some of which are based on caregiving responsibilities that have
been traditionally assigned in a differentiated way. Redistribution of roles among
men and women will take place, but the cultural roots are very deep and the
stakes are very high. . . . Our near term decisions therefore will involve personal
attitudinal change, organizational change, and systemic change. These decisions
themselves are interdependent . . . a set of human resources issues of a scope
larger than we have had to address for many years . . .

Wilson's remarks, delivered in 1991 at the National Academy of Sciences, aroused considerable ire. Some accused her of devaluing the competitiveness of research, which in turn is equated with diluting quality as
detected, sorted, and rewarded through peer-review-based decisionmaking.
Wilson never proposed such a compromise. Rather, some of the problems
facing science "may be linked to the existing ground rules of competition."
"Macho" or "killer" science belongs to an earlier era when men were men,
scientists were men, and science was all about the mastery of nature instead
of harmony among parts of the natural universe. In short, Branscomb's
linear versus ecological thinking.
    Wilson's emphases are not just semantic. The way we talk reflects the
way we think, and often shows how we are inclined to act. Holding the root
values of the research university enterprise up for inspection allows for its
entanglements—planned and unplanned—to be reconsidered. They are not
necessarily wrong, but may be maladaptive for the times, offering models
for behavior to a once homogeneous research community that has become
demographically, ethnically, and intellectually as heterogeneous as any of
us might have ever imagined was possible. We should glory in it and exploit
it, not ignore or suppress it.
    To treat the United States government-science compact as the model
for all time (and all cultures) would continue to stigmatize changes as "problems," "anomalies," and departures from the good old days. As the context
has changed, so has the doing of science and the scientists themselves. The
idea of federally funded research is still sound; its expression (the terms of
discourse and the expectations of participants and interested observers) is
like speaking in tongues. Shared meanings, such as, what is a "partnership"?,
are lost, motives doubted, and disappointments abound.

PROSPECTS FOR A NEW COMPACT

I confess, to write about research and development as a system is to sound alternately preachy, self-righteous, and cynical, or at least disgusted with the system as it is and its portrayal by its dominant participants (Chubin, 1994). This is a time of self-doubt and painful self-examination. It challenges scientists matured by one system to contemplate another—one that is leaner, practical, efficient, fair, and yet uncompromising in its recognition and support of bedrock values: creativity, quality, and risk-taking.

At stake, according to researchers, is the integrity of the research process and a unique national resource known as the university. Politics is viewed as inimical to researchers' interests, with the participation of non-specialists in either the direction or conduct of their research an unnecessary evil. The dichotomy, of course, is false. That is what is meant by a new government-science compact. The need is for new habits of mind; new expectations about federal research funding; new understanding of the word "discretionary"; and a new responsiveness to questions raised by political actors, as coached by their support agencies and policy analysts.

The Reagan '80s, cited as a watershed for research and development funding, also exposed a growing sophistication (or disdain, depending on your viewpoint) about scientists as "just another interest group." It is payback time. The 1990s must demonstrate that the self-interest of scientists is in the best interest of the nation, if not the world. Lurking behind all this is what is quaintly called "the federal role." How should it change? As we contemplate the size and purpose of the United States research and development enterprise, the new compact must achieve these four goals:

Bring People in . . .

The link between fiscal resources and human resources cannot be an afterthought. This is a time to be inclusive, not exclusive. The proliferation of set-aside programs at NSF stigmatizes both the recipients of these competitive, merit-based funds and the rationale for awarding them. If these programs are seen only as adjuncts to the core research funding in disciplines, then they will drive the wedge deeper between national goals and the utilization of talent.

The "winning strategy" for the 1990s is one of participation, and federal agencies must spearhead such affirmative action. Expanding the education pipeline, however, will be wasted motion if career opportunities do not exist or access to them is not vigilantly protected. Not only are women, American minorities, and those with disabilities underrepresented in scientific careers, but they are also underparticipating in the networks of high-level decisionmaking. Women in science advisory circles are particularly scarce. Full participation will require more than adherence to the letter of the law; it will mean fundamental change to the culture of science.

Disabuse Scientists and their Patrons That the University Can Be the Panacea for All the Nation's Ills

We are in the midst of a "shakeout" of universities. Which models will dominate and which values will prevail? Some will strengthen their toe-holds on federal and industrial research resources and remain "research universities." Others will specialize in undergraduate education and select fields of research, development, and graduate training. Some will become centers for statewide and regional technology innovation and economic development.

All will rely on sizable and sustained support from multiple sources, each with its own agenda and influence over administrations, faculty, and students (that is, future workers). Universities will have to decide how best to contribute to national goals-which functions to surrender and which to consolidate. No longer will a single model of competitive research blind universities to the other roles they can perform, and assuredly are playing.

Think Globally: Connect R with D and S with T

All must be connected with other federal policies. The State Department may be more vital to American research and development prospects in the next 10–20 years than the NSF or any of the mission agencies. Because science and technology will not be university-centered, greater appreciation of the melange of sponsors and performers of research, especially the federal laboratories and agency intramural programs, will foster a more balanced policy.

Universities will become more global entities, tied electronically and through international collaborations, than national resources serving single governments and "competitiveness" missions. What is economically competitive will have to enrich many sectors and markets.

To go further, science policy—a small, esoteric specialty—may be an anachronism. Its practitioners need to identify with prominent domestic and international issues and "think bigger." Integrating science and technology policy to achieve national goals is something that policy analysts, comfortable in their professional community niches, must learn to do, indeed transcend.

Nurture Science Watchers and Critics

Arguably and ironically, the cause of public understanding, or at least appreciation, of science has been advanced most strikingly through the scandals and controversies (creationism, cold fusion, animal rights, fraud in research) portrayed on the front pages of our daily newspapers. While negative in content, these reports demystify science and humanize the people and institutions that are making news. This, too, is a form of "popularization," displaying both the integrity and the mischief which a public must observe and criticize to get the whole picture of achievement and accountability in a participatory democracy.

Finally, my bias is that science watchers and critics will be social scientists, by credentials or experience. They have a knack and seasoned eye for translating science and technology. They have a trained capacity for seeing science and technology in society and can relate their interactions to social and economic progress. What the science watcher sees increasingly is that the size the research and development enterprise should be will increasingly depend on how much we know about research and development investments and their performance, and how well we manage the portfolio relative to other federal expenditures in any given year.

Yet will the scientific community commit to reinventing what it has long taken for granted? How inclined will scientists be to enlarge their concept of citizenship, play more active roles in the political system, forge a compact with policymakers and the public that is transparent and bold? The answer to the question "How large a research and development enterprise?" depends on a vision incorporating ecological systemic thinking. Neither a breadth of budget (for none will be robust enough) nor the benevolence of an earlier era conjured by the name Vannevar Bush will rescue American research and development. It is up to all of us scientists. It is our promise to keep.

REFERENCES

Angier, N. (1991). Women swell ranks of science, but remain invisible at the top. *New York Times* May, 21 1991, C1, C12.

Breneman, D. W., and Youn, T. I. K. (eds.) (1988). *Academic Labor Markets and Careers*, Palmer Press, New York.

Chubin, D. E. (1990). Misinformation and the recruitment of students to science. *BioScience* 40:524–526.

Chubin, D. E. (1994). How Large an R&D Enterprise? In Guston, D. and Keniston, K. (eds.), *The Fragile Contract*, MIT Press, Cambridge, MA, pp. 118–144.

Hazen, R. M., and Trefil, J. (1991). General science courses are the key to scientific literacy. *The Chronicle of Higher Education* 10 April: A44.

Heilbron, J. L., and Kevles, D. J. (1989). By failing to discuss the "civics" of science and technology, history textbooks distort the past and endanger the future. *The Chronicle of Higher Education* February 15, 1989, A48.

Heuer, K. (ed.) (1987). *The Lost Notebooks of Loren Eiseley*, Little, Brown, Boston.

How much science is enough? (1991). *Washington Post*, April 24, 1991, p. A20.

Langenberg, D. N. (1991). Science, slogans, and civic duty. *Science* 252:361–363.

Latour, B. (1987). *Science in Action: How to Follow Scientists and Engineers Through Society*, Harvard University Press, Cambridge, MA.

Merida, K. (1994). Americans Want a Direct Say in Political Decision-making, Pollsters Find, *Washington Post*, April 21, 1994, p. A19.

Mullins, N. C., and Chubin, D. E. (1985). Society for social studies of science: perspectives on interdisciplinary research. In Mar, B. et al., *Managing High Technology*, Elsevier, Amsterdam, pp. 175–178.

Oakes, J., Omseth, T., Bell, R., and Camp, P. (1990). *Multiplying Inequalities: The Effects of Race, Social Class, and Tracking on Opportunities to Learn Mathematics and Science*, Rand Corp., Santa Monica, CA.

Pendlebury, S. (1991). From the lab to the tube: Surviving television appearances. *The Scientist* 10 June: 19–20.

Pool, R. (1990). Freshman chemistry was never like this. *Science* 248:157–158.

Prewitt, K. (1982). The public and science policy. *Science, Technology and Human Values* 7:5–14.

Rhodes, F. H. T. (1991). Shaping the future: Science and technology 2030. *Physics Today* 44(5) May 1991: 42–49.

Rigden, J. S., and Tobias, S. (1991). Too often, college-level science is dull as well as difficult. *The Chronicle of Higher Education* March 27, 1991, A52.

Sargent, J., and Pfleeger, J. (1990). The job outlook for college graduates in the year 2000: A 1990 update. *Occupational Outlook Quarterly* 34:3–8.

Schmitt, R. W. (1989). Universities of the future. *Research-Technology Management 32*(5), September–October:18–22.

Sommer, J. (1987). Big three, little four. *American Scientist* 75:103.

Tobias, S. (1992). *Revitalizing Undergraduate Science: Why Some Things Work and Most Don't*, Research Corp., Tucson.

U.S. Congress, Office of Technology Assessment (1985). *Demographic Trends and the Scientific and Engineering Work Force*, U.S. Government Planning Office, Washington, D.C.

U.S. Congress, Office of Technology Assessment (1988). *Elementary and Secondary Education for Science and Engineering*, U.S. Government Printing Office, Washington, D.C.

U.S. Congress, Office of Technology Assessment (1991). *Federally Funded Research: Decisions for a Decade*, U.S. Government Printing Office, Washington, D.C.

Wilson, L. (1992). U.S. universities now confront fateful choices. *The Scientist*, March 16, 1992:11.

CHAPTER 11

Trends and Opportunities in Science and Technology Studies
A View from the National Science Foundation

Edward J. Hackett

The Science and Technology Studies (STS) Program at the National Science Foundation (NSF) spends about $3 million per year to support research and related activities in the history, philosophy, and social studies of science and technology. In conjunction with its close cousin, the Societal Dimensions of Engineering, Science and Technology Program (SDEST), STS supports and guides the field. In this chapter I would like to comment on the field and its future, drawing upon the perspective and experience I acquired during a 2.5 year tour as program director.

In the reflective tradition of STS, I will first mention something about my background and sketch the organization of the NSF and the place of

Edward J. Hackett, Department of Sociology, Arizona State University, Tempe, AZ 85287.

Science, Technology, and Society: A Sourcebook on Research and Practice, edited by Kumar and Chubin, Kluwer Academic / Plenum Publishers, New York, 2000.

the STS program in it. Then I will describe the organizational culture of the NSF, introducing the motif of paradox (or tension between competing values) that will continue throughout the chapter: An overview of ongoing STS projects will follow, giving some impression of the range of work supported by the program. The chapter concludes with some comments about the future of the field.

ORIENTATION

I entered the field of STS while a graduate student at Cornell, by joining the Social Analysis of Science Systems Program and attending the 1976 meetings of the Society for Social Studies of Science on campus that November. I have been in the field, more or less actively, through the intervening decades. I joined the STS Department at Rensselaer Polytechnic Institute in September 1984, served on the advisory panel for the STS program from 1990–1993 (as the first social scientist to advise the "History and Philosophy of Science Program," as it was then called), and left for NSF in the early days of 1996. My predecessor, Ron Overmann, had held the job for some twenty years and deserves all the credit for building a vibrant and diverse program. Ironically, Ron was one of the program officers who funded the large-scale training and research program that supported me as a graduate student.

I do not fit easily into any one STS camp or faction. Most of what I do is empirical, informed by constructivist work but with a healthy dose of critical realism (Hackett, 1990; Chubin and Hackett, 1990). Some of my work is qualitative, some theoretical, some concerned with policy issues. I have never been postmodern.

THE NATIONAL SCIENCE FOUNDATION

The National Science Foundation (NSF) is an independent agency of the federal government, which means it is a part of the government that is not organized within one of the cabinet departments. The NSF spends about $3.5 billion per year to support basic research, education, and related activities in science and engineering. It is organized into seven directorates, one for education and human resources, one for engineering, and one for each of five major fields of science (for example, biological sciences, mathematical and physical sciences). The STS program is located within the Directorate for Social, Behavioral and Economic Sciences (SBE), in a division of Social, Behavioral and Economic Research (SBER). With an annual budget of about \$120 million, the SBE is the smallest directorate—less than half the size of the next largest directorate.

The STS program's neighbors within SBER are more than a dozen programs which represent traditional academic disciplines (such as anthropology, economics, geography, sociology, and the like) as well as interdisciplinary fields of research (such as decision, risk, and management science; law and social science; or measurement, methods, and statistics). STS and SDEST are at the less traditional, more interdisciplinary end of the program continuum.

The STS community must remember that the NSF is a *science* organization, meaning that it *supports the conduct* of science and *promotes the interests* of science. While the field of STS has taken a critical stance toward science and has questioned the epistemological warrants of scientific knowledge—moves that have catalyzed new work in the field and sharpened its intellectual edge—such perspectives are not always consistent with the NSF's mission. At bottom, the activities the NSF supports must contribute to systematic, scientific knowledge either directly (by producing such knowledge) or indirectly (by improving the context for doing so). This organizational mission constrains the sorts of activities the program can support and complicates the interface between the STS program and the field.

The division that houses the STS program takes the "science" aspect of social science quite seriously and for an excellent reason. While STS may justify its presence in the NSF and its expenditure of public science funds either because the work we do *is science or* because the work we do *concerns science*, other SBER programs may use only the first of those two justifications. After all, if these programs are not doing the social *science* of economics (or anthropology or geography or . . .), then why do they belong in the National *Science* Foundation? If instead they are supporting a broader intellectual enterprise—say, "humanistic scholarship"—then perhaps their program belongs in the National Endowment for the *Humanities*.

As STS scholars gnaw at the special epistemological warrants of science, they might think about the consequences of such actions for the funding program that supports graduate training and research for the field.

ORIGINS OF STS

The NSF has supported research in the history and philosophy of science for more than twenty years, chiefly through a program in the History and Philosophy of Science located within the biological sciences directorate. In 1990, at the advice of a Committee of Visitors, the program was renamed

"Science and Technology Studies" and its scope broadened to include "social studies of science and technology." In 1992 the program was moved from the biological sciences directorate to the newly-created social sciences directorate. The new name and new home do not merely reflect organizational changes at the NSF, but they also embody changes in the organization of university departments (Cornell, MIT, and Rensselaer have departments of that name; the University of Pennsylvania has long had a doctoral program in the history and sociology of science; see also Hauger, this volume).

The new name is important in that it signifies fundamental changes within each of the broad fields that intersect in STS. In *history* of science and technology, the move toward STS was catalyzed by a change in analytic framework from internal studies of the sequence of scientific theories or discoveries through external studies that took account of biography, social context, and other influences to integrative work that views the internal–external distinction as contrived and unnecessary (for examples see Cowan, 1983; Hughes, 1983; Star, 1989).

In *philosophy* of science, the move toward a field of STS was catalyzed by a departure from prescriptive philosophy, limited to justification and divorced from practice, to a naturalized philosophy that takes account of history and research practice that is willing to ask questions about the origins of ideas and theories and addresses pressing social or scientific issues (e.g., causation in behavioral genetics or the puzzles of quantum physics). Some examples of such work include Giere (1988) and Longino (1990).

In *social studies* of science and technology, the move into STS is marked by the departure from narrowly structural studies of the roles or careers of scientists or engineers into studies that address the content of scientific or technical work, the social shaping of ideas, designs, and material objects, and the interaction of content with context, organization, and role (for example, Traweek, 1988; Bijker *et al.*, 1987).

The field is young, so it is a bit early to expect an integrated body of theory and research that explains the origin and development of knowledge. The field needs to continue thematic work within the separate strands, each informed by the others, and to try periodically to attain a higher level of integration among strands of research.

Some STS work is provocative—even threatening—to scientists in precisely the same way any systematic study of human social organization, behavior, or cognition may be provocative or threatening. Indeed, it provokes and threatens in the same way that the Apollo missions took away some of the moon's romantic mystery, or genetic research erodes some of the mystery of life, or research on human origins brings with it a certain disenchantment. This critical, questioning stance represents the best of what STS can offer as an intellectual enterprise, but work in this style must be carefully crafted to fit within the framework and mission of the NSF.

THE ORGANIZATIONAL CULTURE OF THE NATIONAL
SCIENCE FOUNDATION

The structural features of the STS program's organizational home—a social science division within an agency which supports and promotes basic research in science and engineering—is not the only force that constrains and shapes the program's decisions. The NSF also has a distinctive organizational culture that influences the way the program goes about its work.

Social science texts traditionally define culture enumeratively as "that complex whole which includes knowledge, belief, art, morals, law, custom . . ." (Tylor, 1871), often summed up by noting that culture provides "a blueprint for living." Within an organization, according to this view, the culture provides a pattern for action, a template for planning, and a vocabulary of legitimate actions. But such definitions—theories of culture, really—overemphasize the common, shared, homogeneous aspect of culture, while downplaying *differences* and *dynamics*.

An alternative view treats culture as a symbolic or moral space, cross-cut by *axes* which represent the value dimensions along which a given culture is positioned and which describe its potential for change (Erikson, 1976). This view gives less weight to the average or central tendencies of a culture and more weight to the variety of values present within a culture and to their internal inconsistencies and capacity for change over time.

The organizational culture of the NSF exhibits several value tensions or inconsistencies, some of which deserve mention here. At its core, NSF is both an *academic* and a *bureaucratic* organization, meaning that it simultaneously honors the traditional academic values—pursue knowledge, reward merit, think and act with originality, share ideas, and so forth—while functioning as part of the federal bureaucracy, honoring all the traditional bureaucratic values as well (such as pursue efficiency, reward behavior that follows established rules, think and act and communicate within the hierarchy).

This has consequences both profound and shallow for how the program operates. Among the more profound consequences are limits on a program officer's decisionmaking authority and constraints on the sorts of documentation needed to support particular actions. There is also some tension between the academic value placed on knowledge as an end in itself, and the bureaucratically-derived value of knowledge as a national resource to be developed in service of other objectives (economic growth health, or security). From an academic standpoint, one would wish to be agile in pursuing new ideas and opportunities, yet from the bureaucratic standpoint accountability and aversion to risk are more salient. So while program officers are often encouraged to take risks, there are others who might later view a risky grant that failed as a waste of tax money.

At the shallow end of the consequence pool are the misperceptions raised when the uninformed read grant titles. One must be careful about the words used when writing their (publicly-searchable) titles and abstracts of funded projects, because one never knows which member of Congress might have their intern search through the database for projects that, to them, sound silly. When such projects are found and pilloried by a member of Congress, NSF responds fully and rapidly, involving the director and deputy director, the Office of Legislative and Public Affairs, the assistant director for the directorate that houses the project (and two or three members of his or her staff), the relevant division director and deputy, and the program officer for the afflicted program. For this reason there are topics and titles and descriptions that may sound fine to the academic ear but would not be suitable for NSF support. Review panels sometimes debate this very point.

A second axis is defined by the *differentiated* character of the organization, with its partition into disciplines, bureaucratic positions, and functions, in contrast to the *integrative*, interdisciplinary, synergistic character of research and education that it tries to sustain. This means that centrifugal and centripetal forces are simultaneously shaping actions and decisions. Such forces arise most sharply in funding decisions or in the formulation of interdisciplinary initiatives, where the shared commitment to the activity that rests at the intersection of fields comes into conflict with the urge to secure as much money as possible for one's own field (an urge, incidentally, that is often reinforced by visits from representatives of scientific societies).

Third, the organization is characterized by both *openness* and *secrecy*. NSF is open to advice about matters general and specific, ranging from the best research opportunities for the next decade to the merits of a particular dissertation proposal. It is also a very permeable organization, with panelists, committees of visitors (that periodically review the management of entire programs), and scientific staff entering from government agencies or academe, then returning home after a one- or two-year tour of duty. It is an organization whose business is to communicate to the outside world of scientists, decisionmakers, students and others. Yet there are also strong rules of confidentiality controlling information about programs or initiatives under development, about decisions that have been made but are not yet ready for release, about reviews and reviewers, about the contents of proposals, and about the very fact a proposal has been submitted.

Fourth, the organization is both *responsive* to new ideas that arise in the scholarly community and *proactive* in creating and stimulating such ideas. To be responsive means a program might collect and evaluate proposals that arise from its research community, allowing the merit review process to decide which proposals to support. Doing this requires little activism on the program officer's part: He or she serves as an honest broker and guide, connecting proposals to reviews and reviews to decisions, with the minimum necessary guidance along the way. But NSF is proactive when it creates an initiative that sets aside a pot of money for new proposals in a new research area, assembling new panels of experts to advise on the decisions. Initiatives are not tailored from whole cloth within the foundation, but arise through an extended process of consultation and advisement. Yet initiatives do draw the research community in new directions, set new challenges, and pose new risks. They originate outside the usual disciplinary frameworks and operate outside the usual NSF programs, managed, for example, by crossdirectorate committees.

Programs and program officers—indeed, the entire foundation—work under the constraints of these normative and value tensions. Beneath the surface language describing the organization's mission, goals and performance objectives (and that implies a unity of purpose and common framework for decisionmaking) is a tumultuous reality of crosscutting considerations and commitments that bear upon any decision. All this, then, forms the organizational stage for the work of the STS program.

HOW STS DECIDES WHAT TO FUND

The STS program provides support for investigator-initiated projects, which means that the idea for a research effort arises from the scholarly community. The STS program announcement, published every two or three years, defines the field in broad terms, then describes several "modes of support," such as fellowships, dissertation improvement awards, training grants, and research grants. Within this framework, specific proposals are written by members of the community and submitted to one of the semi-annual funding cycles.

A proposal arriving at NSF is logged in by the Proposal Processing Unit and given a unique ID number. From there it passed along to the program to be entered into a computer database and considered for possible support. The program officer reads the proposal and identifies six or so outside (or "ad hoc") reviewers, relying on memory, experience, files (e.g., my predecessor kept six card file boxes of potential reviewers, classified by expertise), the literature, membership directories, the proposal's bibliography, and the principal investigator's suggestions to do so.

A review package (consisting of a cover letter, instructions, a review form, and a copy of the proposal) is generated by the program officer and mailed out to each reviewer. Reviewers are asked to evaluate the proposal using two general criteria: What is the intellectual merit of the proposed work, and what is its likely impact? A central computer keeps track of all reviews assigned and all received for at least the past decade. The STS program traditionally uses two stages of review: *ad hoc* reviewers, who are experts chosen specifically to evaluate a particular proposal; and an advisory *panel* that is chosen to review an entire round of proposals, reading the ad hoc reviews, writing reviews of their own, and meeting for a day or two to discuss the proposals.

After the panel meets to advise the program about the merits of the proposals, the program officer does a little housekeeping—for example, ascertains each proposal has received at least three sound reviews—and starts making decisions. Decisions are recorded in a brief document called a "form 7" or "review analysis," which offers a brief justification for the decision to fund or not fund a proposal, drawing liberally from the raw material of the proposal, the reviews (ad hoc and panel), and by the panel discussion. This document becomes part of the permanent record of action taken on the proposal (or "jacket").

Beyond the Mechanics: The Contributions of Peer Review

The STS program relies on reviewers to evaluate proposals and advise decisions, but reviews also serve the STS community in a variety of other ways. For example, reviews provide expert advice to the proposal writer, who receives verbatim copies of reviewers comments. Sometimes such comments can be a little hard to accept—NSF practice is to pass them along verbatim, and no effort is made to resolve inconsistencies in the advice or criticisms they contain. On the whole this practice allows researchers to shape one another's work, and having resources at stake focuses and sharpens the critique. The review process also serves as a flywheel, lending stability and continuity to the research effort. It balances the drive for originality against the inertia of tradition, challenging whether ideas are truly original and whether original ideas are worth pursuing (Chubin and Hackett, 1990). Review also improves communication within the field, sometimes averting duplication (and sometimes spurring competition), and perhaps also preparing receptive ground for new ideas (as they are first presented in a proposal, then substantiated with research and offered for publication).

What's in the Portfolio?

The scope of the program is best seen in a small sampling of projects supported, drawn from the online abstracts maintained by NSF. Anyone wishing more information about these projects or who wants to see what else is underway should visit the NSF web page: www.nsf.gov. The file of project abstracts may be searched by keyword, among other ways. Further information about research programs may be found on program pages.

STS has supported *philosophical* research on classical problems in the philosophy of science, highly technical work in the philosophy of physics, and more topical studies of problems of great current concern. For example, Deborah Mayo is examining the foundations of statistical inference, developing a framework that views experiments as efforts to learn from error, with statistical analysis conceived as a disciplined way to model patterns of irregularity. By weaving together strategies and techniques for learning from error that are employed in fields as diverse as statistics, engineering, physics, biology, and psychology, she is uncovering strikingly similar analytic processes that have been overlooked in many philosophical accounts of scientific inference. J. Richard Creath is examining the philosophical foundations of what it means for a scientific assertion to be "empirically supported." While it is easy to say that assertions gain credibility when they have empirical confirmation, it is hard to explain precisely what such confirmation looks like without slipping into a circular argument. Drawing on newly available works by Rudolf Carnap and W.V.O. Quine, Creath will first develop their competing views on the question of empirical support, then extend those arguments to bear on current issues in the philosophy of science. His study will ultimately provide a reasoned basis for distinguishing between science that rests on empirical justification and "science" that originates from divine or received sources. Helen Longino is concerned with a related problem: reconciling sharp differences between views of science proposed by constructivist and their philosophical opponents. She is developing an analysis that respects the social character of scientific inquiry while leaving room for philosophers' traditional normative concerns about science. Ronald Giere is addressing a similar problem, attempting to bridge the chasm between "scientific realists" and "humanistic relativists" by developing a synthesis that rests on the idea of representation or map-making.

More technical, and far more difficult to describe in accessible terms, are philosophical studies in the foundations of physics, by scholars such as

Robert Batterman, T.Y. Cao, Richard Healey, Paul Teller, and others. Such studies are generally concerned with the philosophical soundness and implications of contemporary physical theories.

A final sort of philosophical work is represented by Kenneth Schaffner's examination of behavioral genetics research in four organisms (worm, fly, mouse, and human). His main concern is to ask what it might mean to claim that a particular gene is responsible for a particular behavior, and to develop a standard for examining the evidence offered for such claims in the scientific literature. With different theories of behavioral genetics establishing different mechanisms and standards of evidence, expert guidance becomes essential for nonscientists to understand and evaluate "discoveries" proclaimed in the media.

Historical research in STS concerns both science and technology, in the ancient, medieval, early modern, modern, and contemporary periods. Most of the work focuses on Europe and the United States, although a small and increasing fraction addresses events occurring outside the western world. History is by far the largest category of research supported by the program, so it is difficult to give more than a superficial sampling of the work underway.

In the history of technology, Jessica Riskin is examining the early 18th-century origins of automation to explore how automatic devices become a foil for our thinking about a range of artistic, biological and philosophical ideas. In the early 20th century context, Emily Thompson is concerned with the interplay of technology, art and business in the development of early sound movies.

In the history of science, Richard Kremer and Michael Shank are delineating how the work of Regiomontanus, the leading mathematical astronomer of the 15th century, may have laid the foundation for Copernicus' revolutionary work. Anita Guerrini is writing an account of "public anatomy," the ceremonial dissection and examination of animals that occurred at many European medical schools during the early modern period. Such public events certainly taught the public some anatomy, but they also entertained, enlightened, and offered moral education as well. Public anatomy is a precursor of animal experimentation and reveals aspects of the change in thinking from religious to secular modes of thought. More difficult to classify is Maurice Finocchiaro's sweeping critical history of literature about the Galileo affair, from its inception in 1633 to the culmination of his "rehabilitation" by the church in 1992. In the 19th-century history of science, Paul Lucier is concerned with geologists and chemists who consulted for the coal and petroleum industries. He is using these cases to consider the role of science and technology in industry, the early development of the norms of scientific practice, and the emergence of the scientific expert witness. In 20th-century science, Lillian Hoddeson is leading an interdisciplinary research team concerned with the genesis and demise of the Superconducting Super Collider as an instance of big science in the contemporary United States. Spencer Weart is writing the history of global climate change research, covering the period 1946–1978, which offers a case where the intrinsic uncertainties of basic science intersect with compelling public concerns.

Social research projects include Michael Fortun's study of the impact of recombinant DNA methods in the field of human genetics and the resulting demand for a concerted effort to map and sequence the entire human genome. Another facet of genetics research and biotechnology concerns Woody Powell and Kenneth Koput, who are examining networks of interorganizational relations among biotechnology firms and university researchers. J. Rogers Hollingsworth and Jerald Hage are using historical data to assess the organizational characteristics that give rise to major discoveries in the life sciences. Other studies compare the politics of biodiversity in Kenya and the United States, the comparative political economy of telecommunications, and the social organization and dynamics of research groups.

STS research uses a wide spectrum of methods, from philosophic analysis that examines premises and the process of drawing inferences from them, through archival and oral history techniques, to the ethnographic and survey tools of social science. In this eclecticism the program serves as a point of contact for disciplines that might otherwise have little to do with one another and offers a strategic site for understanding the sort of interdisciplinary collaboration that may benefit other social sciences.

WHAT LIES AHEAD?

A goal of STS theory is to develop an integrated explanation of the origin, development, shaping, and justification of new knowledge that takes account of ideas, social forces, human behavior, historical dynamics, institutional arrangements, and the material dimensions of science and technology. Tighter integration among STS fields, and stronger links to other disciplines in the social sciences and humanities, will contribute to this goal, enriching all participating parties. Integrated theory is the grail, but it will not come easily because strong institutional forces oppose it. Such forces include the diverse memberships and cultures of professional societies, along with the investments members and officers have in retaining so many distinct associations. Also important are size and resource differences among societies, which influence every effort to cooperate. Finally, it is in the nature of STS thinking to be skeptical of all efforts at intellectual leadership and synthesis, for it is plainly more advantageous to enlist than be enlisted, to synthesize the works of others rather than have one's own work subsumed in another's grand synthesis. In all, reflexivity brings with it some significant baggage.

Beyond integration, the future holds several more focused challenges for the field:

1. It is very important to expand the STS research agenda into other cultures and institutions, thereby broadening the foundation for inference and widening the range for uncovering new ideas. Studies and perspectives from outside the Western context are essential for completeness and comparison. Active encouragement, varied and risky investments of funds, and a ground-up investment in skills through fellowships and training grants are needed to accomplish this. Among the studies that would address this need is Lawrence Schneider's history of experimental genetics in China that covers 1920–1980.

2. The intersection of science and law is an increasingly intense but problematic nexus, which virtually guarantees that it is a rich research site. Competing epistemologies, educational deficiencies (on both sides of the professional divide), novel legal circumstances, and institutional power struggles combine to heighten the tensions between law and science. The stakes are high and rising as more and more science- and technology-based issues are brought to the courts and as judges and jurors (with the assistance of attorneys and expert witnesses) are asked to sort them out (Jasanoff, 1995). Paradoxically, however, the legal system seems less and less prepared to use such evidence in its deliberations: Lawyers deconstruct scientists' testimony; experts battle to a standstill; and judges adjudicate as best they can for the puzzled jury (or step in themselves to sort out the matter).

3. There is an increasing role for STS in science education, which offers a great opportunity for collaboration between STS and the sciences. In the end, STS representation in science textbooks should move from the sidebar to the matrix in which science education is embedded. Discovery and invention are processes, not events, embedded in a continuing stream of human activity and imagination. Students who do not learn this become misled into thinking of science as an activity conducted apart from society by lone heroes for unknown reasons. Not

only is this bad history, but it also sets a bad precedent or pattern for the student's career. Through cooperative educational efforts and through studies of scientific and technical careers, STS can help develop human resources for science, math, and engineering. Of special importance is the possibility that introducing STS issues into the K–12 and post-secondary curricula will show the social relevance of a career in science or engineering and will increase the attractiveness of such careers for all students, but especially for women and members of ethnic groups traditionally underrepresented in those fields.

4. The "activist turn" in STS is the logical next step for the field's commitment to engage issues that are critical or controversial. While this is certainly a valuable turn for the field that merits encouragement, it does not always fit well with the mission of a federal agency that supports science and engineering. The tension between the STS program and its organizational context will not likely abate soon. An imaginative program manager can work with a cooperative investigator to support those aspects of a project that advance STS without conflicting with NSF's mission, and it is important that such collaborations continue. Similarly, STS scholarship should continue to examine the design, implementation, and consequences of information technology (including the Internet) and biotechnology, even though doing so may appear hostile or threatening. With so much at stake, for STS and for society, these lines of inquiry should continue.

5. Electronic media offer unparalleled access to materials and a quantum jump in research possibilities. But this is an expensive and risky route to travel. NSF-wide initiatives in Digital Libraries and in Knowledge and Distributed Intelligence may lead the way, providing cooperative funding with other governmental agencies and "outside the envelope" budget increase for the foundation. Especially noteworthy for STS are program investments in Project Perseus, a web site offering texts, pictures, and site plans of classical Greece; a new effort to build a web site of historical data concerning public health and the development of urban technology in New York City; and the prospect of "digital Darwin," which would make his edited letters available on CD or online. Each effort presents professional research resources to a wider spectrum of users, including remote researchers, students, and scholars engaged in

comparative work. Further, each offers lessons in how to accomplish such dissemination and how to do so more effectively.

6. Finally, both the field and the program should continue to experiment with innovative educational and outreach activities, for students at all levels as well as for the public at large. At the graduate level, for example, the program has provided Small Grants for Training and Research to seed new areas of advanced education. A three-university-consortium used one such award for innovative training at the intersection of history, philosophy, and field work in ecology. Another university group developed a training effort focused on United States science in the Cold War era.

Conferences and workshops are often the seedbeds of new educational and research ideas, for these allow small groups of scholars to work closely together for a brief period of time. One such conference will bring policy makers, activists, scientists, and STS scholars together to examine the research and teaching opportunities offered by a community plagued by various environmental problems. Another will draw upon the history of science to develop educational materials for students in the K–12 grades.

All such activities require risk and imagination in the STS community, and collaboration with and within the foundation. Educational activities are a particularly tender point for social science research programs, because the directorate's budget is dwarfed by that of the education directorate.

CONCLUSION

STS, both field and program, is poised for vigorous growth and innovation. New areas of compelling ignorance have been delineated, and bold scholars are venturing forth. Connections are forming among the constituent fields of STS, between STS and other social sciences, and between STS and the sciences. Educational concerns permeate all of these interfaces; they present opportunities for innovative instruction at all levels, and they offer a more integrative education, which promises to improve dialogue and advance the integration.

Despite such promising possibilities, there are also real and enduring tensions between the field of STS and the STS program, and between the STS program and other activities underway at NSF. The STS critique challenges received wisdom and creates discomfort for scientists and science administrators. Such tensions are strong yet manageable, but are unlikely to disappear soon.

REFERENCES

Bijker, W. E., Hughes, T. P., and Pinch, T. J., eds. (1987). *The Social Construction of Technological Systems*, MIT Press, Cambridge, MA.

Chubin, D. C., and Hackett, E. J. (1990). *Peerless Science: Peer Review and U.S. Science Policy*, State University of New York Press, Albany, N.Y.

Cowan, R. S. (1983). *More Work for Mother*, Basic Books, New York.

Erikson, K. T. (1976). *Everything in Its Path*, Simon and Schuster, New York.

Giere, R. (1988). *Explaining Science*, University of Chicago Press, Chicago, IL.

Hackett, E. J. (1990). Science as a vocation in the 1990s. *Journal of Higher Education* 61:241–278.

Hughes, T. P. (1983). *Networks of Power: Electrification in Western Society, 1880–1930*, Johns Hopkins Press, Baltimore.

Jasanoff, S. S., Markle, G. E., and Petersen, J. E. (1994). *Handbook of Science and Technology Studies*, Sage, Thousand Oaks, CA.

Jasanoff, S. S. (1995). *Science at the Bar*, Harvard University Press, Cambridge, MA.

Longino, H. (1990). *Science as Social Knowledge*, Princeton University Press, Princeton, NJ.

Star, S. L. (1989). *Regions of the Mind: Brain Research and the Quest for Scientific Certainty*, Stanford University Press, Stanford, CA.

Traweek, S. (1998). *Beamtimes and Lifetimes*, Harvard University Press, Cambridge, MA.

Tylor, E. B. (1871). *Primitive Culture*, John Murray, London.

Index